Conference on Future Automotive Technology

Markus Lienkamp (Ed.)

Conference on Future Automotive Technology

Focus Electro Mobility

Springer Vieweg

RESEARCH

Editor
Markus Lienkamp
Garching, Germany

ISBN 978-3-658-01140-6 ISBN 978-3-658-01141-3 (eBook)
DOI 10.1007/978-3-658-01141-3

The Deutsche Nationalbibliothek lists this publication in the Deutsche Nationalbibliografie; detailed bibliographic data are available in the Internet at http://dnb.d-nb.de.

Library of Congress Control Number: 2012954490

Springer Vieweg

Printed on acid-free paper

Springer Vieweg is a brand of Springer DE.
Springer DE is part of Springer Science+Business Media.
www.springer-vieweg.de

Preface

The increasing trend towards electric cars issues several challenges for the automobile industry, research institutions and politics as well as for society.

Research and serial development move closer together to meet automotive standards with new components like traction batteries integrated into hybrid and electrical drivetrains. Furthermore, the influence of e-mobility on the daily mobility behavior, the effects on the automotive supply chain and the impact on industrial production have to be taken into account.

According to these complex aspects it is crucial to not only acquire specific knowledge in the particular fields but also to consider their functional interaction. Therefore it seems essential to merge competence from science, economy and politics.

This year, the annual „Conference on Future Automotive Technology“ as the follow-up of the “2. Automobiltechnisches Kolloquium München” sets the focus on the economical realization of widespread automotive electro mobility.

The book at hand contains the papers of the five topics *Energy storage technologies*, *Mobility and Service*, *Powertrain*, *Vehicle concepts and engineering* and *Production* which were presented at the conference.

The energy storage system is the crucial key factor to electro mobility. Low capacity which is directly related to a small range as well as high costs hinder electric vehicles to penetrate into the market. Thus new technologies for range extenders such as zinc-air-batteries and additional flywheel range-extending systems are discussed in this book according to low utilization ratio, reasonableness and efficiency.

In the field of *Mobility and Service* the paper discusses an alternative approach to a clean and affordable mobility for electric vehicles.

The four papers of the topic *Powertrain* controvert the components motor, clutch and transmission.

Especially the design and the operational behavior of a special electric motor as well as the electro mechanically actuated clutch for an optimal friction torque control is discussed. Due to the high rotational speed of electric motors the role of transmissions has to be reconsidered. Thus the need of transmission was reviewed on the conference.

Considering the hybrid electric vehicles (HEV) one paper outlines the possibilities of the powertrain to increase the range of HEV in terms of downsized conventional powertrains and hybrid combustion engines.

Especially the needs and boundaries of electrified cars require new procedures to find the suiting vehicle concepts. Therefore the impacts of electrification on the Vehicle Concept as well as the effects of regional requirements are discussed. Being a pivotal part of the vehicle, the drivetrain offers new possibilities for the architecture of electrified components. One paper discusses the possibilities of free distribution of torque and its effects on the driving dynamics.

Prof. Dr.-Ing. Markus Lienkamp
Institute of Automotive Engineering
Technische Universität München
Garching

Table of Contents

I Energy Storage Technologies

1 Alternative Range Extender for Electric Cars – Zinc Air Batteries

Richard Eckl, Peter Burda, Armin Foerg, Harald Finke,
Prof. Dr.-Ing. Markus Lienkamp

Abstract: The small range of electric cars compared to cars with internal combustion engine is still one of the problems inhibiting the spread of these cars. Range extenders (RE) in general are the realistic possibility to increase the range of electric cars in the near future without the need to improve the technology of the main battery storage. This paper describes an alternative type of RE for electric cars. Its aim is to enlarge the range in special situations and not for the daily use. A study [1] shows that a range of 100 km per day is suitable for 90 % of the daily trips of a year in Germany. The electric car MUTE is designed to work as regional used car mainly for driving to work and for shopping. Range extension for a regional used car is necessary in special situations like driving around road works or traffic jams, which is the aim of this RE. Utilizing the zinc-air technology as RE makes a fully electric car with zero emission feasible. This contribution shows that a zinc air RE with an energy content of 4 kWh and a peak power of 4 kW is feasible to increase the range of regional used electric cars even if the RE is activated at a state of charge (SOC) of the main battery which is below 10 %.

Keywords: Range Extender, Zinc-Air Battery, MUTE, BEV.

1.1 Introduction

The small range of electric cars compared to cars with internal combustion engine is still one of the problems inhibiting the spread of these cars. For this purpose, small internal combustion engines and fuel cells are currently in discussion as range extender (RE). In general, it is also possible to power the electric engine only by RE. Thus it is possible to get almost unlimited range just by refuelling, but this is not necessary in 90 % of the days of a year in Germany as published in a study [1]. For this study 100 km of range per day is enough for the car drivers, especially when the car is used for commuting and shopping. In cases they definitely need more range, for example on weekends for holiday trips, they have to use other means of transportation. This can be a second car which is available for the user. The driver of a regional used car having 100 km of range only might need more range in special situations that could not be planned before starting a trip, e.g. to drive around road works. But this is only necessary few times a year, as we expect. Hence, the seldom use of a RE would be sufficient and reliable during the car life. Furthermore, it would be preferred to become totally independent from fossil fuels also for the RE. This can be fulfilled by zinc-air batteries.

Today, zinc air batteries are typically used as power supply for hearing aids because of their high specific energy, which is about 1.5 times higher than the one of lithium-ion batteries [2]. Their application as main energy storage device in electric cars was investigated by several research facilities in the past. Only few are named here. In 1991 a Honda CRX was converted into an electric car powered by zinc air batteries [3]. It had a range of 346 km and reached a top speed of 140 km/h. For acceleration a nickel-cadmium battery was used parallel to the zinc air battery to provide more power for short times. The zinc air battery used in this case was electric rechargeable up to 75 times.

A refuelable zinc air battery was also tested in an electric shuttle bus [4]. The bus had a range up to 158 km. This zinc-air battery was rechargeable mechanically by exchanging the materials of the zinc anode without disassembling the battery.

Unfortunately the research about zinc air batteries in automotive applications was stopped for some reasons. One reason can be the fact that it is only possible to recharge electrically a zinc-air battery up to 150 times [5], which is not sufficient for an electric car if the battery is used as main energy storage. Also the second possibility to recharge the batteries mechanically seemed not to be feasible for this purpose.

Although zinc-air batteries seem not to be applicable as main energy storage, they still could be useful as electric RE, which is suggested in this paper. Especially, while regarding the study about electric transportation [1] this kind of RE could be attractive. Thinking about a seldom necessity of use up to 10 times a year, an electric rechargeable RE could be used for about 15 years. Regarding this seldom usage, also the mechanic rechargeable RE could be interesting. In this contribution an exchange of discharged battery modules is suggested. The RE should consist out of several modules which can be exchanged manually. The electric car MUTE is already prepared to carry six zinc-air battery modules. It was developed at the Technical University of Munich in collaboration of several institutes. This car can later be used to investigate the ability for this technology as RE. The modules are intended for exchanging at a service station while the discharged ones should be sent back to the factory for refilling. The most expensive part of the battery, the air-electrode, and also the housing could be utilized again. Only the zinc anode has to be changed. The refilled battery modules can be reused as RE.

In the electric car MUTE the RE is placed in the front part of it in the collapsible zone. This can be done because zinc air batteries are short circuit proof, as described in [6]. This prevents the battery from fire and explosion.

1.2 Alternative range extender

1.2.1 Approach

For the analysis shown in this contribution, measurements of zinc air test cells were done at first. Out of the measurement results, a first RE for the electric car MUTE was derived. Several properties can be estimated in this way. For example, it is possible to calculate the total range of the electric car MUTE in case the main energy storage and the RE both are employed to power the engine. The RE is connected parallel to the main energy storage by a DC/DC converter. Therefore it is necessary to distinguish different starting times of the RE at different states of charge (SOC) of the main battery.

The main approach of this contribution is to derive a RE for MUTE out of the measurement data of zinc air batteries and not from theoretical data taken from zinc air technologies. Thus the results of this investigation will be more realistic regarding the state of art in zinc air technology.

1.2.2 Experiments

1.2.2.1 Experimental setup

For the experiments test cell bodies were developed to be able to house different kinds of anodes and cathodes in order to analyse them within the same cell body and thus within the same environment. The test cell bodies made out of polyamide can be reused for several experiments and also to examine other battery components in the future. For these experiments the cell components (Teflon layer, air electrode, separator, anode material) of a Duracell® DA146 were utilized.

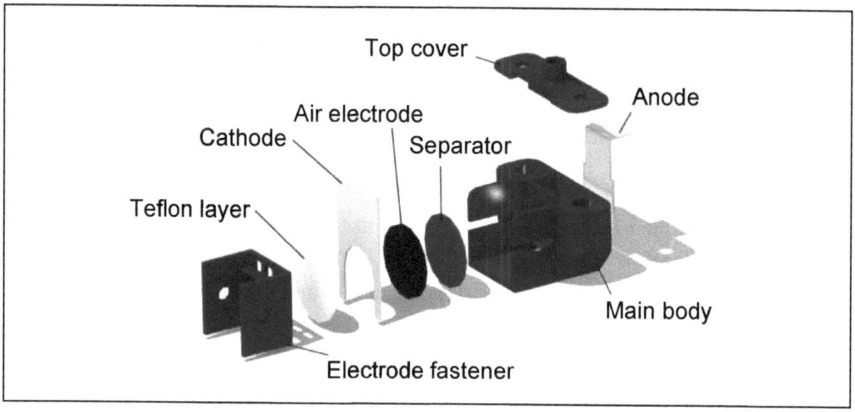

Figure 1: Assembly of the test cell body

The composition of the anode material thus is the same as in a DA146. Figure 1 shows the assembly of the test cell body. It is possible to insert in this cell different air electrodes (cathodes) and different mixtures of the zinc anode to examine them.

The test cell body provides a hole for the air electrode in a way that every inserted air electrode has a resulting active surface of 0.95 cm². The volume of the anode reservoir in the main body is 0.81 cm³. In our experiments 9.9 g of anode material (mixture of electrolyte and zinc powder) were filled into this reservoir. The filled test cells were measured by the battery test system BasyTec CTS LAB XL with measuring board G750 from GAMRY Instruments at the Institute for Electrical Energy Storage Technology. Therefore a test program has been defined. This starts with 3 s of measurement with time steps of 20 ms to get a high resolution of the behaviour at the beginning, followed by 300 s of measurement with time steps of 1 s. Then the measurement period is changed to time steps of 20 s until a voltage of less than 0.01 V is measured, to get the whole discharge curve.

The discharge was done with constant current while the voltage and the energy content of the test cell were measured. The experiments were done with different discharge currents (10 mA, 20 mA, 30 mA, 40 mA, 50 mA and 60 mA). The discharge currents stayed constant over the whole experiment with a refilled test cell. For every discharge current step three test cells were used and refilled for the next step with new battery components.

1.2.2.2 Results

The discharge curves show in general the expected shape of discharging a zinc air battery. Figure 2 shows the discharge of a zinc air test cell at 10 mA of constant current while Figure 3 shows the discharge at 60 mA. The measurements at currents between 10 mA and 60 mA step by step show the expected transformation between both curves. The curve changes from a shape with a flat discharge interval ending with a visible sharp bend to a shape where a sharp bend is not visible any more. According to the step by step increase of the discharge current, the discharge time of the test cells decreased in the expected way, which means doubling the discharge current results in about half of the discharge time.

In the next step the measured energy content of the test cells were compared. Therefore the average energy content has been calculated out of every three test cells which were discharged with the same constant current. The comparison is diagrammed in Figure 4. There the average energy content of every three test cells with the same discharge current and their standard deviation is visible. Obviously a strong correlation between the discharge current and the usable energy content exists for zinc air batteries. If the discharge current increases, the usable energy content of a battery decreases. The measurements show that at a discharge current of 10 mA of a test cell, the energy content is nearly double the energy content of a test cell discharged with 60 mA.

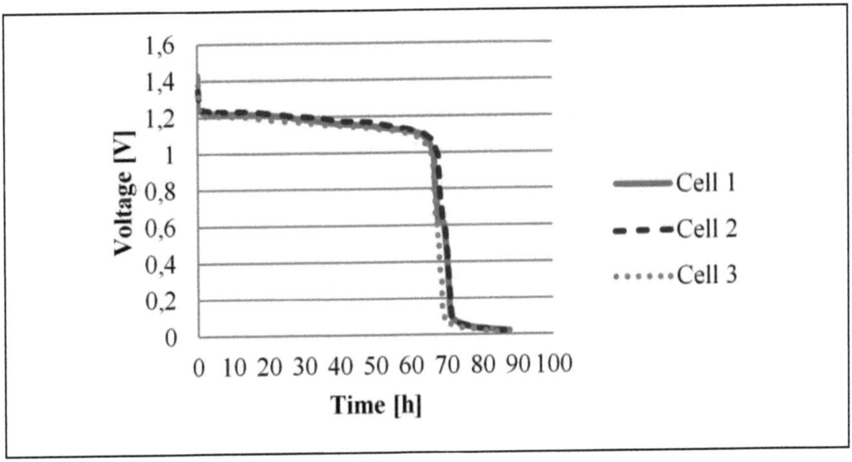

Figure 2: Discharge with 10 mA

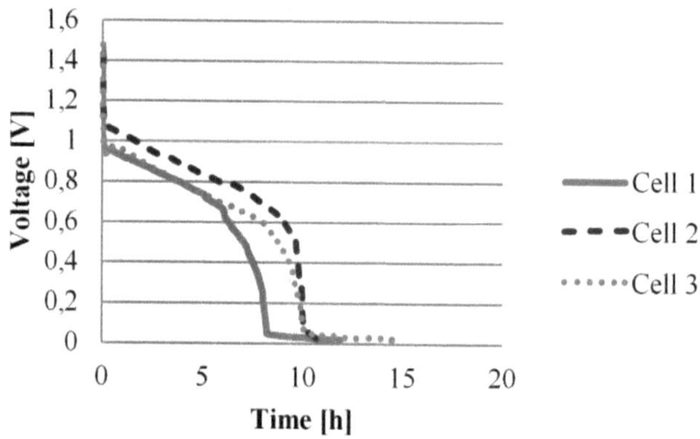

Figure 3: Discharge with 60 mA

This correlation in general is known for a long time for different types of batteries, but it varies from battery type to battery type. As described in [7], the difference between lead-acid batteries and lithium-ion batteries is significant. While lead acid batteries show a strong correlation between the usable energy content and the discharge rate, lithium-ion batteries can attenuate this effect due to a rise of temperature in the cell during discharge. The rising temperature leads to a higher usable energy content, which can attenuate the known effect of decreasing the usable energy content due to higher discharge rates. The same reference [7] shows that the energy content measured in Ah varies only between 30.4 Ah and 30.8 Ah for the analysed lithium-ion battery for discharge currents of 5 and 50 A. Taking into consideration that the voltage of the measured cell during discharge for the two cases is not already accounted for the capacity expressed as energy content in Wh, as described in this contribution, a larger difference between a discharge with 5 A and a discharge with 50 A is obvious. Calculating the usable energy content in Wh out of the Figure 4 in [7], the battery discharged with 5 A has a usable energy content of 102 Wh while the one discharged with 50 A has a usable energy content of 94 Wh. Thus the energy content of lithium-ion batteries does not vary as much as it does for the zinc air batteries presented in this contribution. Furthermore, the ratio between the specific energies of lithium-ion batteries and zinc air batteries varies for different discharge rates.

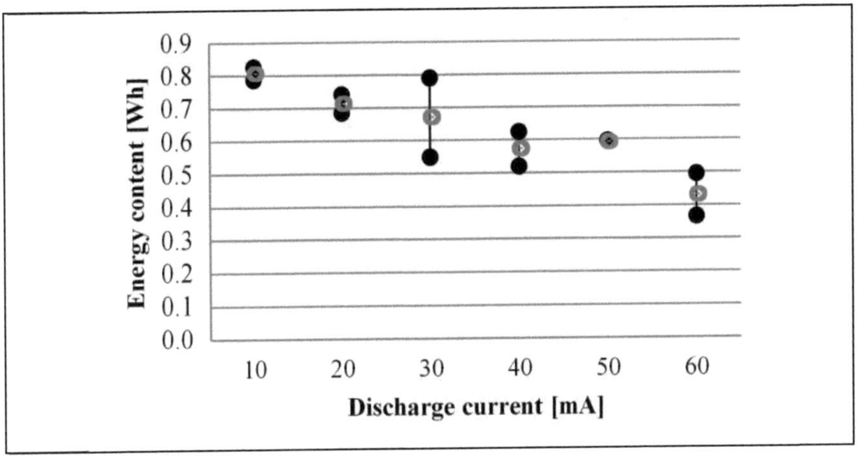

Figure 4: Relation between discharge current and energy content

1.2.3 Calculations

Out of the measurements shown below, some of the properties of a RE were calculated. This has been done for the RE planned for the electric car MUTE. This car has a main energy storage device (lithium-ion battery) with an energy content of 10 kWh. The RE is planned to have an energy content of 4 kWh. To be able to replace the discharged mechanic rechargeable RE, it consists out of six modules containing 36 zinc air cells each. This design leads to a weight of each module that eases manual handling. Having 36 cells in series, each module reaches a voltage level up to 50.4 V in case of a maximum voltage of 1.4 V of every cell. All modules are connected in parallel. Thus, the touch voltage of the RE is still below 60 V. This is necessary for special operating conditions of electric systems which can be touched by a person [8]. The RE is connected to the high voltage system of the car through a DC/DC converter. It is possible to charge the main energy storage and to drive the electric motor by the RE or partly by the RE together with the main energy storage. Especially for accelerating phases, the output power of the RE is too low to feed the electric motor on its own. Hence, an energy management is required, since it is necessary to have some energy left in the main energy storage to be able to drive the car by RE.

Figure 5 shows one RE module consisting out of 36 cells, a handle at the front side and the air duct on the top. The arrows show the air convection into and out of the module which is forced by fan. The air flows from the inlet into the

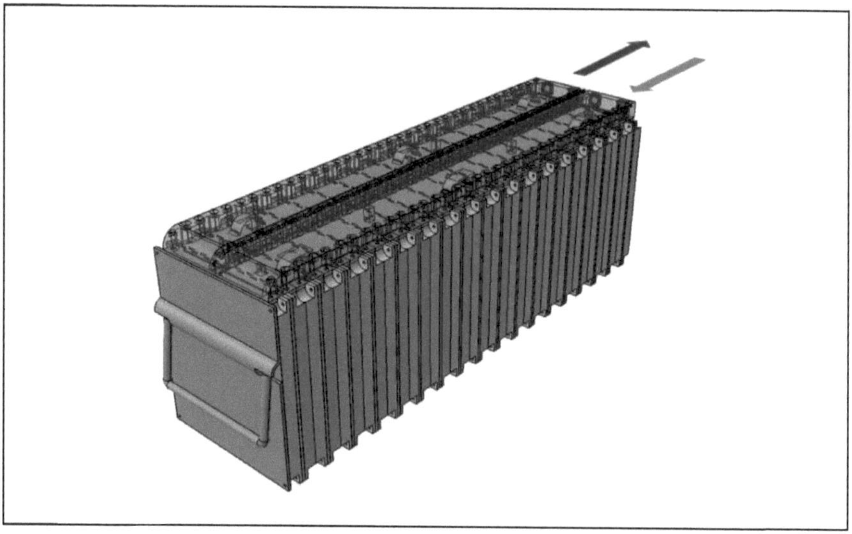

Figure 5: One RE module for the electric car MUTE

module and then downwards to each cell. It forms a u-shape and comes up again to the air duct on the top of the cells. Then the air flow leaves the module at the outlet at the back side.

1.2.3.1 Range Extender for MUTE

It is possible to derive a suitable RE from the measurement data. Assuming that we planned an energy content of 4 kWh for the RE with a peak power of at least 4 kW into the intended space in the front part of the electric car, the resulting weight of the required anode material is 40 kg. Furthermore the RE is required to fit into the front part of the electric car. This RE works with discharge currents comparable to those measured in the test cells, namely 60 mA, proportionately to the active surface of the air electrodes.

In general, the specific energy of zinc air batteries is about 1.5 times higher than the one of lithium-ion batteries [2]. By deriving different REs from the measured data, it is evident that this is not true for every output power. Corresponding discharge rates of a zinc air RE were shown in 2.2.2 Results. Figure 6 shows the relation of the maximum output power to the usable energy content of possible REs. The real specific energy content decreases while the maximum output power increases.

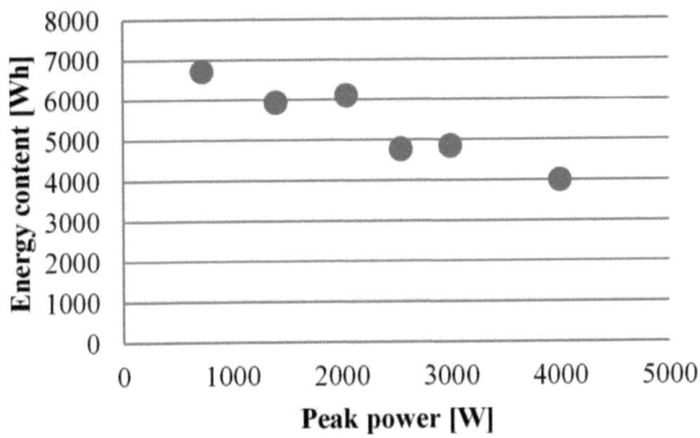

Figure 6: Relation between peak power and energy content

The same relation can obviously be found for the average output power to the energy content as shown in Figure 7. For the calculated RE with 4 kW peak power an average power of 3 kW was calculated out of the discharge curve of the derived RE shown in Figure 8.

The curves shown in the Figures 6 and 7 are derived from the curves calculated for the different REs having an equivalent discharge current like the test

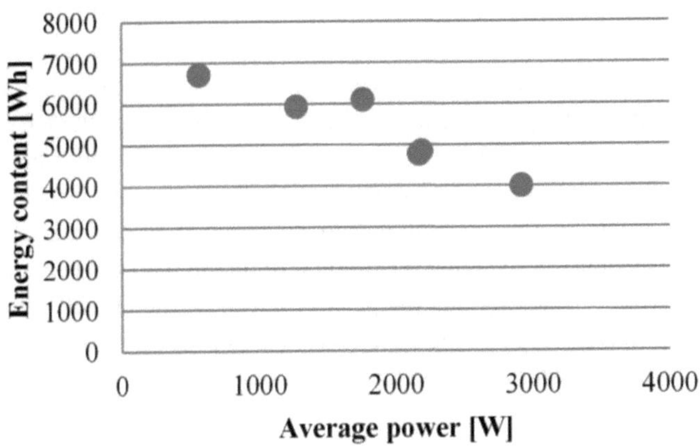

Figure 7: Relation between average power and energy content

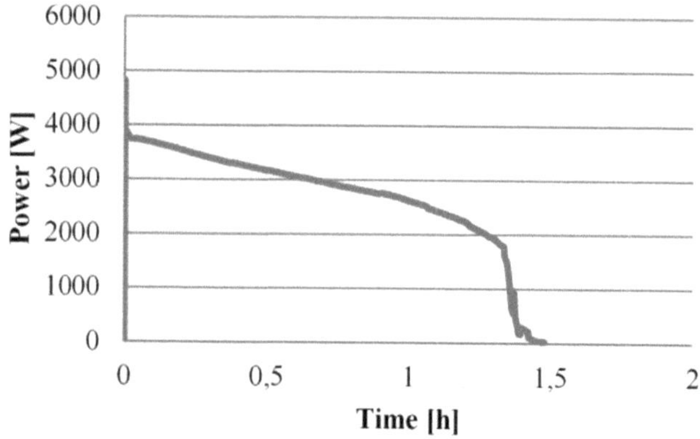

Figure 8: Discharge curve of a possible RE for MUTE

cells from the measurements, proportionately to the active surface of the air electrodes. In Figure 8 the discharge curve for the calculated RE with 4 kW peak power and 4 kWh energy content is shown. The shape of the discharge curve is comparable to the best one of the test cells with a discharge current of 60 mA. The curve of the RE was calculated using this measured data as starting point. In the calculation, the cell size was scaled up. The active surface of the air electrode of one RE cell was changed from 0.95 cm^2 of the test cells to 260 cm^2 of one MUTE RE cell. Those cells are planned to have a height of 140 mm, a width of 109 mm and a thickness of 10 mm. Hence the air electrode should be placed on the front side and the back side of the cell to get enough surface for the air electrode and thus to reach the aimed peak power.

By scaling up the energy content together with the weight of the RE cell, from the test cells it is evident that the energy content of the RE would be much too high because the scaling up was done proportionally to the air electrode surface. Consequently, such energy content has to be reduced to the aimed value of 4 kWh. This was done by reducing the amount of anode material per surface area of the air electrode. Assuming that the peak power of a zinc air cell is mainly depending on the surface area of the air electrode this procedure is feasible. Thus, the amount of anode material can be varied without affecting the peak power.

For this first analysis of an application of zinc air batteries as RE it is not necessary to distinguish between mechanical and electrical recharging. Both are feasible. Nevertheless, electric recharging would have the advantages that the battery could stay in the car and be recharged. The mechanically rechargeable battery has the advantage that a lot of the knowledge from cells for hearing aids can be used, but during mechanic recharge of the test cells the problem of the discharged anode material showing a solid consistency occurred. It was not possible to remove the discharged material easily.

1.2.3.2 Range of MUTE

To analyse if the suggested RE is reasonable for the application in MUTE, calculations with driving cycles were made. In a first step a Matlab/Simulink simulation model for MUTE was used. For the simulations a possible recuperation up to a breaking acceleration of -1 m/s² was assumed.

The driving cycles used for this analysis are the New European Driving Cycle (NEDC), the Common Artemis Driving Cycle (CADC) and the FTP72. For the calculation of the range of MUTE using the main energy storage together with the RE, average energy consumptions were used which were calculated with the Matlab/Simulink model for the different driving cycles. The range afterwards was calculated separately from the average energy consumption. Besides the range itself, a very interesting point is the starting time of the RE. It can be activated at different state of charge (SOC) of the main energy storage. This allows us to predict if the RE with the planned peak power is reasonable in a car like MUTE. The RE should be able to enlarge the range even if the SOC of the main battery is very low, to be reasonable. In most of the use cases, when drivers need additional range it is expected, that they remark this at a lately stage of time. Thus, it is of special interest to consider the starting time of the RE at a SOC of the main energy storage below 10 %. Figures 9 to 12 show the maximum range of MUTE using the main energy storage together with the RE while it was activated at different SOC of the main battery.

Figure 9 shows the range of MUTE depending on the starting time of the RE for the NEDC. Obviously the RE can enlarge the range for starting times below 10 % SOC significantly. Only for SOC below 1 % the range becomes shorter than the maximum range.

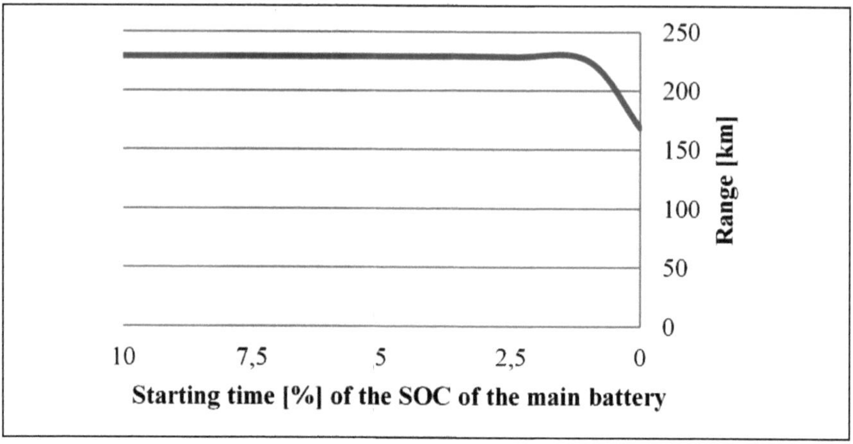

Figure 9: Range of MUTE calculated with NEDC

To cover the case that MUTE is used mainly in the suburban area with extra urban driving conditions, a calculation for this part of NEDC was done separately. Figure 10 shows the range of MUTE in extra urban conditions. In this case, the reduction of the maximum range starts earlier at starting times around 30 % SOC, but at 2.5 % SOC it is still possible to get 7.8 km of extra range out of the RE. In the Graphs, which show the maximum range depending on the starting time of the RE, at 0 % SOC of the main battery always the range without RE is visible.

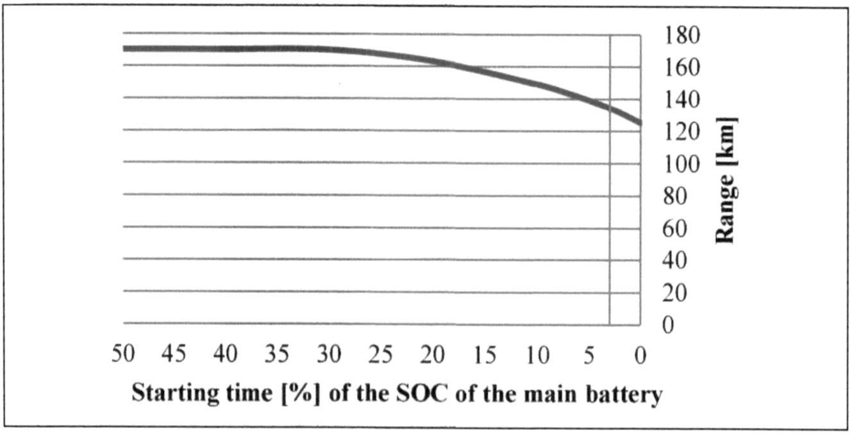

Figure 10: Range of MUTE calculated with EUDC

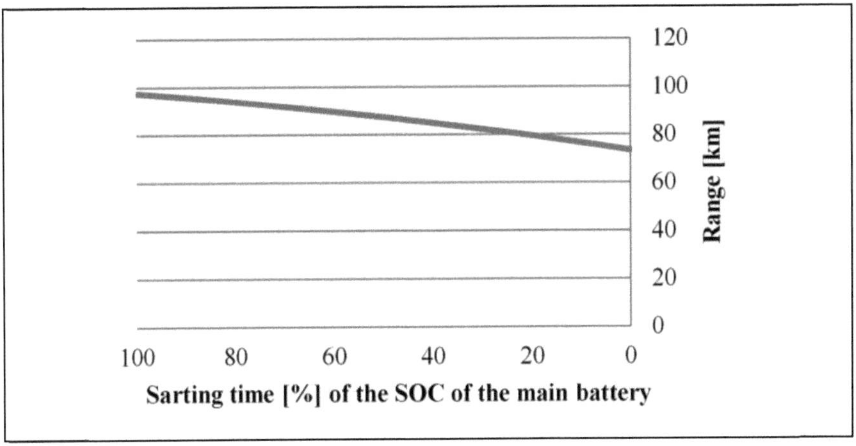

Figure 11: Range of MUTE calculated with the complete CADC

The CADC is divided into three sections (urban, rural and motorway). MUTE, due to its top speed, is only able to drive through the urban and the rural cycle. Nevertheless, also the range for the whole CADC has been calculated for MUTE and is shown in Figure 11. As expected this curve shows a totally different shape than the others. MUTE is not able to drive at the predetermined speeds of the CADC motorway cycle. Thus MUTE will not reach the energy consumption calculated for this part of CADC. Figure 12 shows the range for the CADC urban

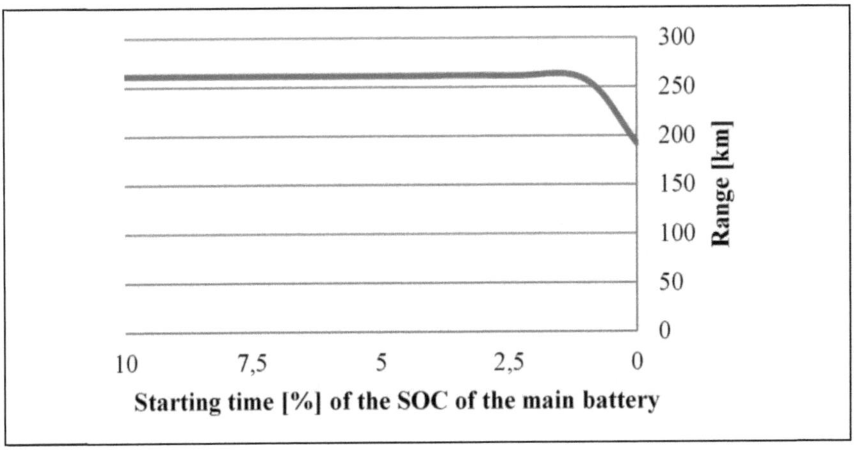

Figure 12: Range of MUTE calculated with CADC urban and rural

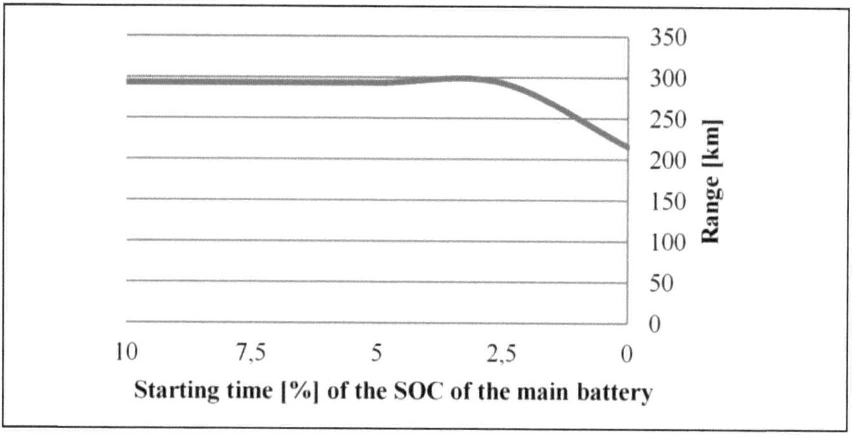

Figure 13: Range of MUTE calculated with FTP72

and rural. The shape of the curve is nearly the same than in the NEDC but the range is 30 km higher which is maybe due to the missing CADC motorway part.

As driving cycle for urban conditions the FTP72 is established in automotive industry. The range calculated therefore is shown in Figure 13. The range stays constant until the starting time of the RE around 2.5 % SOC of the main battery.

As known from practical experience, the driving cycles often show too high values for the possible range. Thus, these results also might tend to too high values, but they show that, even at low SOC of the main battery, a RE with 4 kW peak power and a usable energy content of 4 kWh is able to enlarge the range of the electric car MUTE for seldom use.

1.3 Conclusion

Measurements on zinc air test cells were done for this contribution. Based on the results, it was possible to extrapolate the working properties of a zinc air RE from the available measured data. This contribution shows that it is possible to enlarge the total range of an electric car like MUTE by a zinc air RE having 4 kW peak power, and 4 kWh of energy content between 45 and 77 km for the standard driving cycles. This range extender includes in total 40 kg of anode material. Even at a SOC of the main battery less than 10 % it is possible to enlarge the total range of the car by activating the RE. The most critical driving cycle showed that at a SOC of 2.5 % it is still possible to get 7.8 km of extra range. All analysed driving cycles that make sense for MUTE show that this is possible.

On the other hand, obviously there are still some open points left which have to be solved. Due to the measurements with different discharge currents, it is noteworthy that the usable energy content of zinc air batteries in practical use is much lower than expected. The advantage of higher energy density compared to lithium-ion batteries is questionable for high discharge rates of zinc air batteries. The mechanical recharge in our application was handicapped by the solid consistency of the discharged anode material. It was difficult to remove it.

1.4 Acknowledgements

My thanks go to the Institute for Electrical Energy Storage Technology, Munich, Germany for providing the measurement technology used for this research and to the Institute for Technical Electrochemistry, Garching, Germany, in particular to Michele Piana, for the support concerning the chemistry of zinc-air batteries and in reviewing the manuscript.

1.5 References

[1] Boecker, J. et al. “VDE-Studie Elektrofahrzeuge”, VDE Verband der Elektrotechnik Elektronik Informationstechnik e.V., Frankfurt am Main, Germany, April 2010, p 5.

[2] Dixon, J. “Energy Storage for Electric Vehicles”, IEEE International Conference on Industrial Technology (ICIT), Viña del Mar - Valparaiso, Chile, March 2010.

[3] Danczyk, L.G.; Scheffler, R.L.; Hobbs, R.S. “A High Performance Zinc-Air Powered Electric Vehicle”, Future Transportation Technology Conference & Exposition, Portland, OR, USA, August 1991.

[4] Cooper, J.F. et al. “A Refuelable Zinc/Air Battery for Fleet Electric Vehicle Propulsion”, Future Transportation Technology Conference and Exposition, Costa Mesa, California, USA, August 1995.

[5] Haas, O. et al. “Metal/air batteries: The zinc/air case”, in Handbook of Fuel Cells – Fundamentals, Technology and Applications, Volume 1, Part 4, John Wiley & Sons, Ltd, Chichester, 2003, pp 382–408.

[6] Cheiky, M.C. et al. "Rechargeable Zinc-Air Batteries in Electric Vehicle Applications", Future Transportation Technology Conference and Exposition, San Diego, California, USA, August 1990.

[7] Doerffel, D.; Sharkh, S.A. "A critical review of using the Peukert equation for determining the remaining capacity of lead-acid and lithium-ion batteries", Journal of Power Sources 155, 2006, pp. 395–400.

[8] Kiefer, G. "VDE 0100 und die Praxis", 13. Auflage, VDE Verlag GmbH, Berlin / Offenbach, 2009, p.104.

2 Electric Car Operation and Flywheel Energy Storage

Prof. Dr. Andreas Daberkow, Marcus Ehlert, Dominik Kaise

Abstract: Since 2009 Heilbronn University has been investigating the specific needs of individual and commuter traffic for electric car operation in urban-regional areas. The plug-in battery-powered university research car has 26 lithium-based batteries, each with a capacity of 160 Ah and a voltage of 3.2 V. From approx. 11000 km of test drives with measurements and recordings of state variables like motor current or regenerative braking current it becomes clear that the available range allows reliable operation not only in urban, but also in mixed urban-regional areas. However, range remains an issue so that further research was started on additional flywheel range-extending systems. The paper reports first results of the flywheel system investigations. With a flywheel operation speed of 40 000 rpm basic effects of energy regeneration are investigated. Also, first results of the investigations concerning flywheel dynamics on the vehicle dynamics are presented.

Keywords: Electro mobility, Flywheel Energy Storage, Battery Electric Vehicle Operation

2.1 Introduction

The regional confederation Heilbronn-Franken is typical for a large number of mixed urban-regional districts in Germany and also in Europe, too. Typical for the Heilbronn-Franken area close to the southwest metropolitan area of Stuttgart is a slightly hilly landscape as seen in Figure 1.

Electro mobility in the understanding of battery-powered road vehicles promotes pollution reduction and energy saving in transportation. Heilbronn University was supported in 2009 by the ZIP (future investment program) program of the Baden-Württemberg state to investigate electrically powered passenger car impacts on mixed urban-regional transportation (initial project funding).

Figure 1: Typical landscape view of South-west regional district of Heilbronn-Franken [1] and University Electric car during winter overnight charging

Intense discussions with electric car manufacturers in 2009 led to the procurement of a CITYSAX vehicle [2]. In close discussions with the manufacturer and the local service partner Auto Neff Heilbronn this Chevrolet Matiz conversion vehicle was supplemented by measurement and recording equipment. Figure 2 shows 8 of 26 traction batteries located in the space of the former engine cooling unit. The asynchronous electric motor with a permanent power of 16 kW and a peak power of 26 kW is adapted to the 5-speed gearbox clutch that is adopted from the gasoline-driven model. On top of the front traction battery unit and the electric motor a platform holds the 220 V DC to AC charging device, the AC to DC motor converter equipment and the 83.2 V DC to low voltage 12 V DC converter.

Further batteries are located in the former spare wheel compartment of the vehicle, battery charging and overall control units are located in the inner

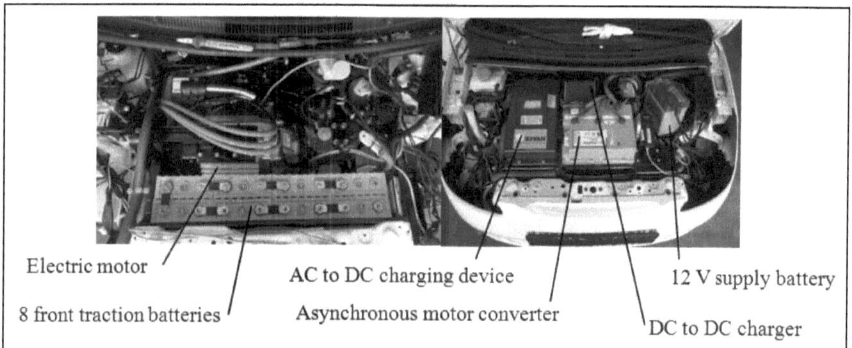

Figure 2: Electric car and components, front car view

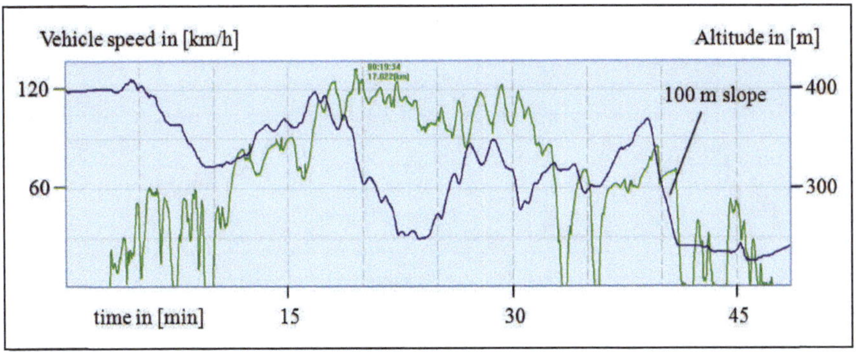

Figure 3: Vehicle speed and GPS logging of topographic profile of electric vehicle in 50 km driving cycle

compartment of the vehicle. The rear cargo volume provides space for additional equipment like fast charging or future range-extending prototype devices.

Now, the Heilbronn University team has tested more than 11000 km with the CITYSAX electric car in autumn, winter and spring conditions, see [3]. Even in cold winter conditions of 2012 (including partially overnight parking with no weather protection) an operation was reliable, due to the gasoline-driven parking heater the internal compartment heating had no effects on range. It was decided to choose a set of realistic urban-regional driving cycles divided into ranges of 5, 12, 25, 50 and 75 km. The 50 km driving cycle is a one-way driving scenario, the reverse direction of 50 km is passed after recharging the car. Figure 3 shows the topographic profile and the vehicle speed in the 50 km cycle from Stuttgart to Heilbronn, Figure 4 shows the velocity and the power consumption from data loggings in the reverse direction.

It is remarkable that a peak power of 26 kW is sufficient to compete in today's high-speed motorway traffic. Also, Figure 4 shows the energy regeneration advantage by braking indicated by negative power consumption values.

The Heilbronn University team is convinced that electric car operation in mixed urban – regional areas is feasible for commuter traffic today. However, range is an issue for electric car operation today and in the future. Therefore range-extending techniques were investigated immediately and in close relation to the test driving cycles.

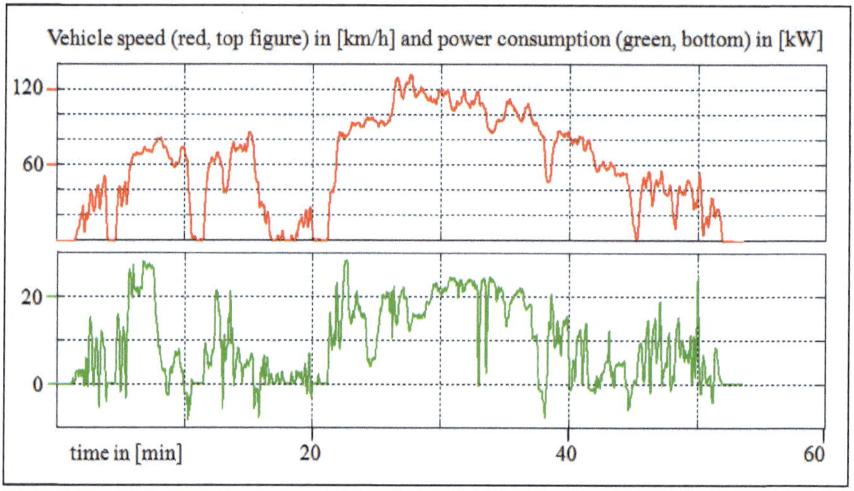

Figure 4: Vehicle speed and power consumption in 50 km cycle, opposite direction

2.2 Flywheel range extending approach

Range-extending techniques for electric vehicles are in lively discussion today. One of the first-ever series plug-in hybrid cars available today is the Chevrolet Volt unveiled in 2007 by the General Motors Company. The Chevrolet Volt and the similar model Opel Ampera or Vauxhall is called a plug-in hybrid electric vehicle that can be operated as a purely battery electric vehicle. When its battery capacity reaches a defined discharge state, its gasoline combustion piston engine powers an electric generator to extend the vehicle's range, see e.g. [4]. Further range-extending approaches are based on the Wankel rotary engine or on super-capacitors. The range-extending scenario described here is based on the flywheel energy storage system principle. Flywheel energy storage and propulsion systems have been investigated in some industrial applications, primarily for stationary emergency energy supply or for the delivery of high energy rates within a short time period. Figure 5 shows the Gyrobus as an automotive application that was built and operated 1953 in Switzerland. Its flywheel had a weight of 1.5 tons and revolved with a maximum speed of 3000 1/min, see [5].

Figure 5: Gyrobus [5]

New flywheel materials like carbon fiber and further progress in high speed power electronics led to a renaissance of the flywheel approach. Recent deployments of flywheel storage systems are described in Porsche race technology applications, see [6]. Here the flywheel which is connected to an electric drive system rotates with a maximum speed of 40000 1/min. Also the Volvo Motor company announces an additional flywheel storage system with a maximum speed of 60000 1/min, see [7] and Figure 6.

There are several advantages in a flywheel energy storage system. On the one hand, its basically mechanical assembly allows multiple discharges and recharges that are not possible for modern lithium batteries. Also the short-term delivery of high energy rates is not compatible with a long duration of battery life. On the other hand, high speed flywheels need additional devices for the reduction of air and bearing friction that enlarges self-discharge. Also, the installation within a mobile platform may cause vibrations and may need appropriate suspension devices, see [8].

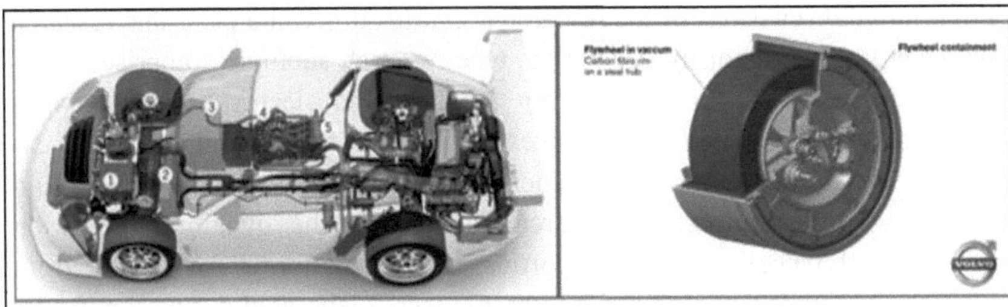

Figure 6: Porsche (left) and Volvo (right) flywheel energy storage application descriptions from [6] and [7]

In the scope of this research project two basic scenarios of preview range extending and fast recharging are considered that differ from today's range-extending strategies.

Preview range extending

In this case the Range extension via flywheel energy storage is supposed to be done before starting a car ride for which more energy is needed. For this purpose, the flywheel range extender is charged on demand and its energy is consumed at the beginning of the ride. During this phase the flywheel range extender saves traction battery energy. This also takes into account of the fact that the flywheel energy may be needed for trip specific compartment heating, compartment cooling, peak acceleration or extreme uphill slope requirements. This preview operation mode takes into account that nowadays an average continuous battery electric car ride does not exceed 1-2 hours. It is designed to prevent flywheel power demands in the case of disadvantageous self-discharge conditions.

Fast recharging

In this operation mode range, extension via flywheel energy storage is supposed to be available when an unexpected recharge due to trip conditions is necessary. Here, the flywheel principle is advantageous since fast charging reduces the lifespan of today's lithium-type batteries. A mid-term goal is an emergency recharging suitable for an additional 20 min or 20 km electric car trip.

For both cases it has to be decided if energy regeneration via braking is transformed into the flywheel or the traction battery capacity. Further flywheel energy storage applications may include emergency power supply for a vehicle's electric network balance or even for stabilizing the ride comfort.

From initial discussions of the research scenarios it became clear that research activities are advantageously decoupled. One research subtopic comprises questions about energy conversion between flywheel and electric motor/generator unit including power electronics devices. All results from simulations and experiments for this subtopic were generated in the Master study project and thesis [9]. Another important research subtopic is the dynamic interaction between the flywheel and the supporting mobile platform. All results from simulations and experiments for this subtopic were elaborated during the master study project and thesis [10] and will be explained in the next chapter.

2.3 Flywheel and mobile platform interaction

After theoretical prestudies on flywheels, see e.g. [11] it was decided to establish a simulation model for mobile platforms close to the University Electric Vehicle design. Therefore, the mobile platform simulation model is provided with an integrated flywheel unit with a maximum of two revolving disks. All model setups and simulations were performed with the computer simulation tools MATLAB/ Simulink and SIMPACK. A proper validation of the simulation model and its results was an important goal of the project. Initial validation plans for a new smaller scale laboratory mobile flywheel design and manufacturing were discussed and abandoned due to project time and costs. Instead of this, the research project team decided to follow the practical idea described in [10]. Here, the mobile platform testbed is simulated by the programmable movements of an articulated industrial robot available in the Heilbronn University robot laboratory. It allows the indication and reproduction of typical yaw and move motions to a versatile adapter that connects with a smaller scale flywheel device. The connector itself includes a torque and force measurement unit to identify the internal forces and torques applied during the enforced robot motion. Finally, a commercial camera flywheel stabilizing device from Keynon company [12] was chosen and mounted to the connector, see Figure 7.

The flywheel device consists of two hinged flywheels with external power supply and a maximum speed of 20 000 1/min rotating in a helium-tight housing.

Comprehensive work was necessary on the set-up, calibration and validation of this approach. Since an internal design of the cardanic hinges and force elements of the flywheel device was not documented, intensive work including X-Ray video recordings was required to identify the flywheel device parameters

Figure 7: Mobile platform testbed with articulated robot, connector and flywheel device

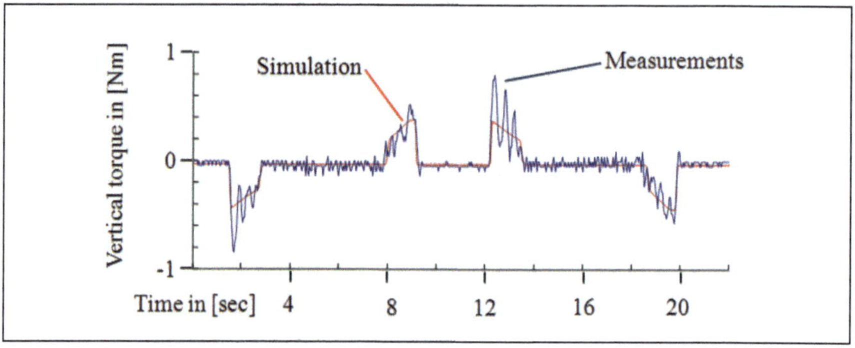

Figure 8: Vertical torque of flywheel stabilization device during yaw motion

itself. Figure 8 shows the vertical torque from measurements and simulations induced from a yaw movement with 24°/sec. Due to the internal elasticity of the force and torque sensor, all measurements were superimposed by vibrations.

After these preceeding investigations a validated simulation model was set up in Simulink and SIMPACK and finally ready for adaption and scaling to a mobile platform. This platform is a first approximation with the parameters of the subcompact electric vehicle. The simulation model now allows variable parameter and hinge design modifications as well as different flywheel speeds, orientations and moments of inertia, see Figure 9.

First simulations show the influence of flywheel hinging modifications on the vehicle movements including stabilizing effects. Further simulation and analysis work on the basis of this validated and scalable simulation model is planned.

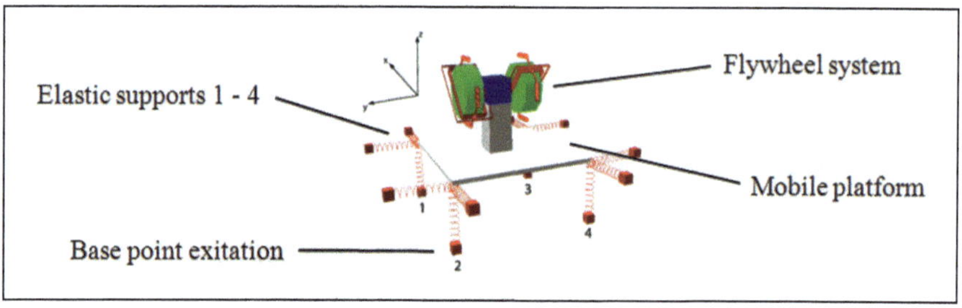

Figure 9: Simulation model of flywheel on mobile platform

2.4 Energy Dynamics

All basic work concerning the interaction between flywheel, power electronics and the electric motor/generator unit is comprised in the research study [9]. Together with the power electronics company ARADEX GmbH and the flywheel manufacturing company Rosetta GmbH a stationary flywheel system was simulated, designed, manufactured, installed and tested, see Figure 10. All simulations were done with MATLAB/Simulink.

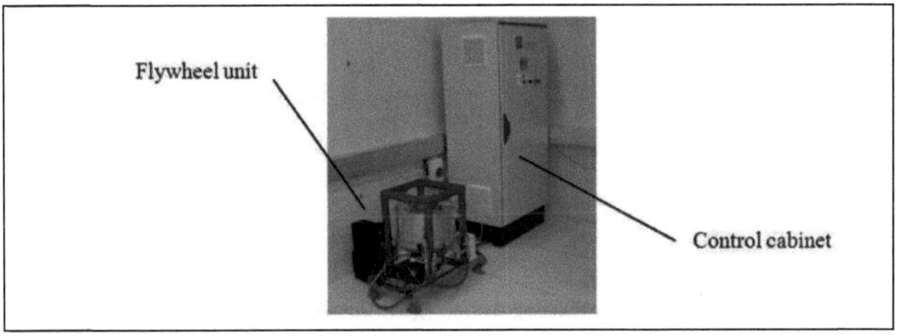

Figure 10: Stationary flywheel testbed with flywheel unit and control cabinet

The simplified laboratory flywheel prototype without carbon fibre wheel body has a maximum speed of approx. 40 000 1/min and an energy content of 70 kWs. To reduce complexity no vacuum was generated, the energy was consumed via electric recovery into the University electric supply network. The double-pole synchronous electric motor/generator unit has a peak power of 15 kW.

Parameters for the moments of inertia and especially for bearing friction were calculated and measured in close collaboration with the flywheel supplier. The characteristic curves for friction torque modeling are determined from deceleration tests. Comprehensive tests including high speed acceleration, constant speed runs and fast deceleration show a very good compliance between simulation model and measurements, see Figure 11.

This testbed is now designed for flywheel devices with a higher energy content and higher speed. It also became clear that self-discharge is one of the main optimization targets for flywheel energy storage systems. Based on the simulations and measurements, a flywheel model with an energy content of 1.5 kWh was simulated together with the University electric car and Heilbronn drive cycle parameters. A theoretical drive length of 8 min including the ascending slope of

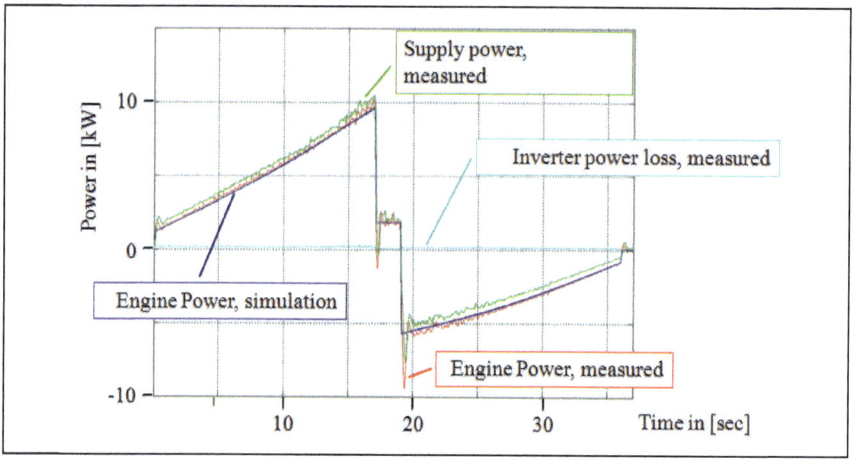

Figure 11: Flywheel power characteristics from fast acceleration and deceleration within 35 s

Figure 3 shows that further work must be done on the fly-wheel layout and especially on the bearing and air friction. In the next step the energy recovery into the University electric supply network will be replaced by a controllable resistance energy consumption.

2.5 Conclusion

From 11000 km test experience with a 26 kW powered electric car of 13 kWh range, the Heilbronn University team is convinced that electric car operation in mixed urban – regional areas is feasible today especially for commuter traffic. However, range is and will be an issue for all battery electric vehicles. In this paper, a fly-wheel range-extending approach is investigated by theoretical, simulation and laboratory research.

Therefore, two new basic operation strategies are proposed. A "preview range extending" strategy consumes the precharged flywheel energy content early at the beginning of a travel not only for traction but also for heating and cooling purposes. This strategy compensates the effects of flywheel energy losses due to friction. The "fast recharging" strategy holds if unexpected range extensions occur. Here, the flywheel capability of multiple fast charging without reduction of lifetime is utilized. During further initial work, two basic research tracks for the flywheel dynamics are revealed, the flywheel mobile platform interaction track and the flywheel drive dynamics research track.

For investigation of the mobile platform interaction a flywheel camera stabilization device in combination with an articulated robot was utilized. It serves to validate a scalable simulation model now available for a flywheel system with multiple hinge and wheel configurations. For further conceptual work on the university vehicle with supplemented flywheel, typical vehicle motions like yaw and roll and its interaction can be investigated.

A stationary 70 kWs flywheel prototype with 15 kW motor/generator and power electronics unit has been designed in the flywheel drive dynamics research track. In parallel, a simulation model was established and validated with flywheel experiments up to 40 000 1/min. This scalable simulation model is now available to transfer multiple flywheel and motor/generator parameters for electric car adaption. First simulations including a first scaling with a flywheel energy content of 1.5 kWh shows that approximately 15 % of the typical urban-regional test cycles can be powered by the flywheel system.

Further perennial research work based on these results will be supported by the state of Baden-Württemberg. It will focus on the configuration of flywheel mass, speed, diameter and hinge configuration in electric vehicle application at minimum friction costs.

2.6 Acknowledgements

The authors want to thank Prof. Rainer Uhler, Heilbronn University who provides his valuable assistance concerning the modeling and simulation of the controlled electric motor/engine and its power electronics. Also we thank Prof. Andreas Hoch, Heilbronn University for making the articulated robot of the robot laboratory available to us and providing us with his valuable advice and suggestions.

2.7 References

[1] http://commons.wikimedia.org/wiki/File:Neckarbecken_Heilbronn_v_ Michaelsberg _2 0061118.jpg. Referenced March 30th, 2011.

[2] Heyne, D. "Electric Car CITYSAX: Two Saxons design their own E-Car" Auto, Motor + Sport, October 2010.

[3] Daberkow, A.; Häussler, S. "Electric Car operation in mixed urban – regional areas", Proceedings of the 13th EAEC European Automotive Congress, Valencia, June 2011.

[4] Scoltok, J. "Opel-Vauxhall Ampera" Automotive Engineer, November 2011.

[5] http://www.fbw.ch/galerie/Gyrobus/gyrotechnik.HTM. Referenced March 19th, 2012.

[6] Hrachowy, F. "Schwäbischer Brummkreisel. Porsche setzt beim Hybrid auf alternative Energiespeicherung", Krafthand 8, Band 83/2010.

[7] http://www.volvocars.com/de/top/about/news_events/pages/press.aspx?itemid=145. Referenced March 19th, 2012.

[8] Daberkow, A.; Häussler, S.; Ehlert, M.; Kaiser, D. „Verfahren und Vorrichtung zur Lagerung eines Schwungrad-Energiespeichers", patent pending, 2011.

[9] Kaiser, D. „Konzeption und Untersuchung eines Schwungradspeichers für elektromobile Anwendungen" Master Study Thesis, Heilbronn University, 2011.

[10] Ehlert, M. "Untersuchung der Dynamik schnellaufender Rotoren auf elektromobilen Systemen" Master Study Thesis, Heilbronn University, 2011.

[11] Magnus, K. „Der Kreisel", Industrie-Druck GmbH Verlag Göttingen, 1965.

[12] Keynon, T. "Stabilizing Instruments" United States Patent Office, 1961.

II Mobility and Service

3 Sustainable Mobility – The MUTE as a Prime Example for Clean and Affordable Mobility

Bodo Gohla-Neudecker, Christian Kandler, Thomas Hamacher

Abstract: Mobility is a key challenge for sustainable development. Mobility can be blamed for an increasing world-wide oil demand, poor air quality in cities and accelerated CO_2 emissions. Electric vehicles seem to offer a way out of these problems, if electricity can be supplied in a sustainable fashion. What is the situation if the current German generation mix is assumed, what if the loading patterns are accounted for properly. In an extreme view a new electricity supply system only for cars is developed based on renewable energies. The integration of an electric vehicle into the house energy system mainly driven by solar power might offer another possibility. These ideas and related questions are discussed in the following article. The MUTE is taken as a prime example to characterise electric vehicles.

Keywords: sustainable mobility, renewable energy, electric vehicle, power plant emissions, charging costs.

3.1 Introduction

Electric vehicles (EVs) are currently a widely discussed and very popular scientific topic. Many expectations rest on their economic and ecological impact on the automobile sector and society as a whole. One crucial factor for the EVs' future success lies in the type of power plants generating their electricity requirements. Effectively EVs' operation costs and attributed emissions are determined herewith. It is thus vital to screen the power production costs and CO_2 emissions of power plants closely, which – next to the general grid load – must provide the additional electricity needed to charge potentially several million EVs.

TUM's MUTE can be considered a prime example for a series of highly efficient EVs designated by some OEMs for the next coming years. They represent subcompact cars with low vehicle masses, highly efficient electric power trains and benchmark rolling and air resistances. Thus for this paper the MUTE's technical specifications will be taken as exemplary input values for simulation purposes in order to quantify and analyse two sustainability criteria (CO_2 emissions and charging costs) for small compact EVs.

This paper will help provide answers to the following relevant and partly still un-answered questions: Can small compact EVs such as the MUTE compete on economic and ecological aspects with their conventional counterparts with internal combustion engines (ICE) and will EVs mark the turning point to a more sustainable future – a future in which conflicting objectives of a cleaner and at the same time affordable individual mobility are attainable any time soon?

3.2 Methodologies for Charging EVs

3.2.1 EV Charging Times and Impacted Power Plants

When examining the impact of EVs on power generation, it is important to consider the hour of day, in which they will be drawing electricity from the grid. To help derive this Figure 1 depicts the average distribution of road trips, which are started at a certain hour during the day in Germany [1].

In many other countries such comparable distributions are common as well, showing the following typical characteristics: Traffic density peaks twice over a period of 24 hours (morning and late afternoon), while just before and especially after midnight barely any trips are commenced at all. As is common perception, traffic jams in densely populated areas arise in the morning, when most commuters leave home for work. And conversely the second traffic peak arises in the late afternoon once most commuters take up their journey to head back home or before that to run errands, go to after-work socials or carry on with other activeties. In between these two peaks, it can generally be established that traffic remains moderate during the day, while at night, due to human inactivity, traffic dies down completely.

Mobility statistics have shown that vehicles on average are parked 22 hours per day without any further use [1]. For EVs this stationary time period would not be any different; yet more relevant to know is the following: During which hours are EVs effectively connected to the grid and thus would be able to charge their batteries? Due to the fact that no statistical or theoretical data on EV grid connectivity rates exists, an assumption is made on when EVs with partly or fully depleted batteries can begin their charging processes. This assumption is

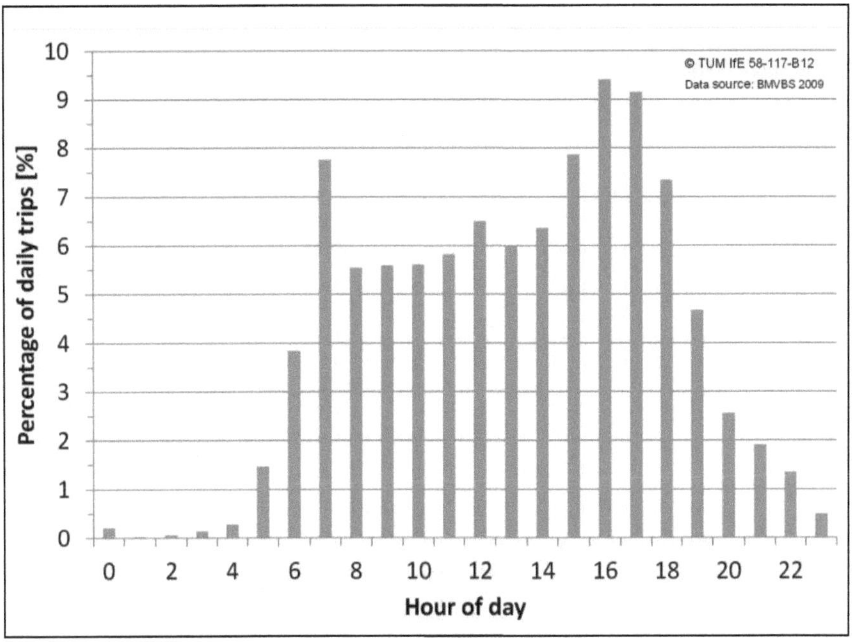

Figure 1: Distribution of weekday road trips in Germany 2008

based on an inverse correlation to the statistical data presented already in Figure 1 on commenced road trips in Germany. Figure 2 presents this assumption, defined as the rate of EV grid connectivity for 2015 (red line) and its possible development until 2050 (blue line).

The question now remains, why grid connectivity is assumed to be lowest during the day at 20 % for 2015: It can be expected, that commuters mainly have the chance of connecting their EVs to the grid at home. Thus only few commuters have the chance to connect to the grid at their place of work or in a designated parking space while running errands or at other daily activities. It is therefore assumed that during the second daily traffic peak, only 20 % of EVs have the possibility to use a charging spot. However, as soon as EV charging infrastructure becomes more common in future, this percentage will increase further and further and may reach its maximum of approx. 90,5 % (direct inverse of Figure 1) once charging infrastructure is assumed ubiquitous by 2050.

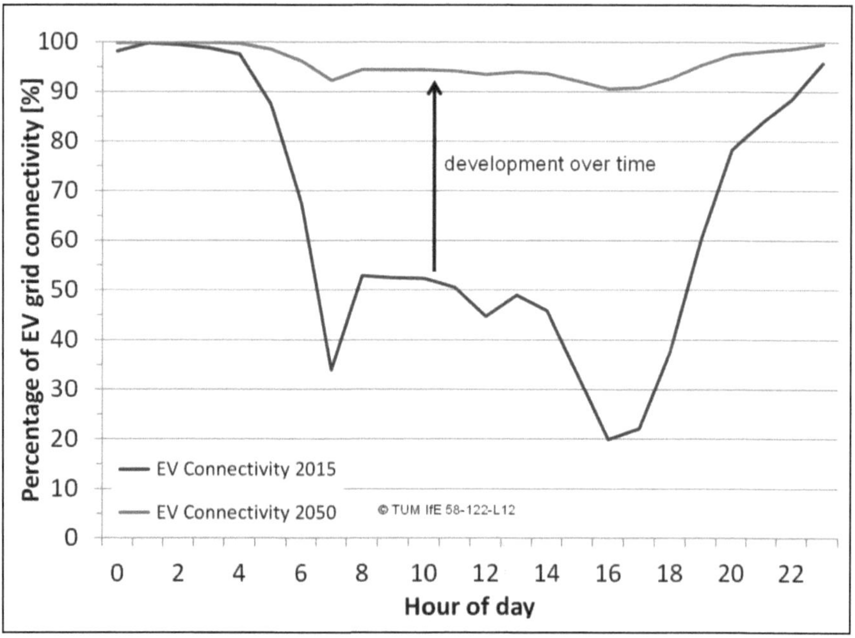

Figure 2: Weekday distribution of EV grid connectivity and its development from 2015 to 2050

The next step in assessing the impact of EVs on power generation is to determine, when EVs actually charge their depleted batteries. While Figure 2 only depicts the times of a physical connection between EVs and the electrical grid, no comments were made on when the vehicles actually draw power from the grid and effectively power plants are required to ramp up electricity production to cater for this additional electricity demand.

The charging times of several EVs can either be aggregated and co-ordinated by centrally located grid management systems or self-controlled by each individual EV user with their own smart meters. This is termed as controlled charging of EVs. If, however, the batteries are charged as soon as the vehicles physically connect to the grid, the term uncontrolled charging will be used. This would then directly correspond to the EVs' usage patterns in Figure 1 and their corresponding grid connectivity in Figure 2.

For controlled EV charging, however, the time to draw power from the grid can be chosen decisively according to certain cost or emission criteria. For example this can be done during off-peak load hours (just before and especially after midnight), when the residual load of thermal power plants is generally lower. This for example is already the case, when subscribing to night time tariffs in order to minimise electricity costs. According to Figure 2 almost 100 % of EVs are physically connected at this time to the grid and would be waiting to receive the go-ahead signal for charging their depleted batteries.

In both cases – uncontrolled and controlled EV charging – different power plants ramp up their electricity production to cater for a higher grid load due to EVs. To demonstrate which types of power plants are impacted, Figure 3 will explain the methodology behind this using the example of controlled EV charging during off-peak hours.

Given in Figure 3 is an average daily load forecast for a typical winter's day. In order to cover the load forecast, power generation capacities are utilised according to the merit order – the ranking of power plants' variable costs in ascending order – so that those with the lowest variable costs are the first to be brought online, while those with the highest variable costs are the last. With barely any operational costs renewable sources are the first to cover the load, followed by base load power plants (nuclear, lignite) and middle load power plants (hard coal, some gas). If the load is almost fully covered by now, such as during the hours just after midnight, all remaining power plants attain either a partial load state or remain idle. Should consumer load increase strongly, as is the case over midday or the second load peak in the late afternoon, additionally peak load power plants based on gas and oil take up electricity generation.

At all times the prerequisite for a stable grid frequency and voltage requires power supply and demand to remain in a state of continuous sync. This is due to the fact that electricity as such cannot be stored. In a comparison between power supply and classical demand, the supply side is by far more flexible than the demand side. However, due to their 22 hours of theoretical grid connectivity, EVs are a game changer in this respect and are able to adapt their charging times very flexibly.

In case of shifting the power demand of EVs into off-peak times, the energy economic term of "valley filling" is used to describe this effect of levelling the load curve (Figure 3). Hereby middle load power plants in partial load or in idle status increase their power generation beyond the general load forecast (grey columns above load forecast curve) in order to provide the additional electricity demand of EVs (light purple columns). This effect of load levelling can be amplified further by means of "peak shaving" (dark purple columns), in which potential EVs feed power back into the grid at times of high load forecasts. This

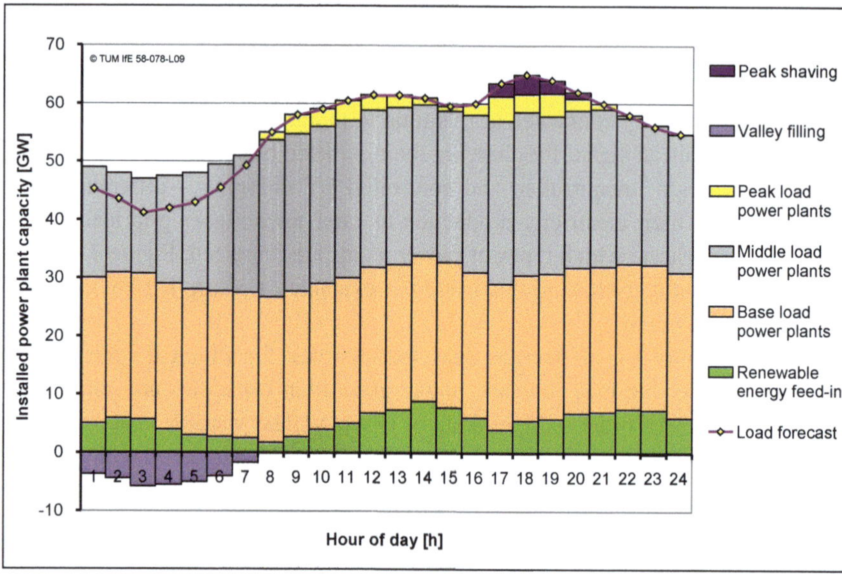

Figure 3: Utilisation of the Merit Order for covering the load curve and additional EV power demand during off-peak hours

effect is also shown in the diagram but is not considered for further discussion in this paper.

When considering uncontrolled EV charging (in correspondence to Figure 2), it can now be deduced that peak load power plants during the day and late afternoon and middle load power plants in the hours before midnight ramp up their power generation to cover the additional electricity demand from EVs.

The EV parameters described above flow into the power plant upgrade and operation model MOWIKA, which simulates the cost optimal future development of the power plant infrastructure in Germany, while considering the impact of EVs as major electricity consumers. Further details on input variables, which encompass technical, economic and political parameters of power plants, fuel prices, technological developments, etc., as well as the optimisation methodology (linear programming) are described in [2].

Now that EV charging times and impacted power plants can be determined, three previously presented methodologies are available for allocating emissions and costs to EVs: the Mix Method, widely used in public discussions, the Delta Method, first presented in [3] and finally the Parallel Market Method from [4]. In the following paragraphs the Mix, Delta and Parallel Market Methods will shortly be recapped and applied to the MUTE.

3.2.2 *The Mix Method for EV Impact Assessments*

When applying the Mix Method, EVs are treated as every other regular electricity consumer. The methodology for assessing the impact of EVs is depicted in Figure 4. Hereby power generation efficiencies, CO_2 emissions and electricity costs are shared by all consumers on the same calculation basis, irrespective of their times of power demand from the grid. Thus the total amount of CO_2 emissions and power production costs are equally divided among the generated kilowatt hours of the whole electricity grid of a country. The average grid mix on emissions and generation costs can hence be established. These are specific values in gCO_2/kWh and in €/kWh. The absolute amount of emissions and costs is hereby determined by the final energy demand of each individual consumer.

This allocation methodology has the effect that the impact of EVs on the operation and investment pattern of the power generation sector is passed on to

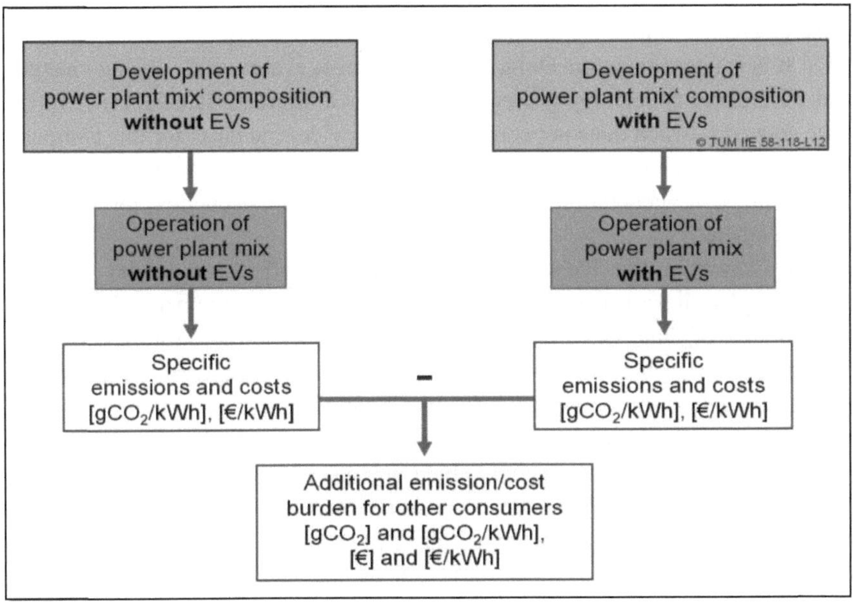

Figure 4: Methodology for allocating and evaluating EV emissions and costs: the Mix Method

all other electricity consumers equally (Figure 4). For a better understanding, this will now be explained in an exemplary scenario for Germany, in which EVs are charged during off-peak hours (controlled EV charging). In comparison to a scenario without EVs, the whole power supply sector emits noticeably more CO_2 emissions and generates electricity at lower costs. This is due to the longer full load hours of mainly base and middle load power plants, which in Germany's case would be lignite and hard coal fired power plants. Therefore in comparison to a scenario without EVs, all electricity consumers incur higher specific CO_2 emissions and lower specific electricity costs.

According to the Mix Method it can be generalised, that EVs shift their primary energy demand and all associated emissions and costs from the transport sector of an economy (mainly crude oil and a sparse amount of biomass as is the case in Germany) to the power supply sector (diverse mix of energy carriers such as nuclear, coal, gas, renewables, etc.). Thus when speaking of decreasing a country's dependency on crude oil by diversifying its primary energy demand with the help of EVs, the Mix Method is the basis for discussion.

3.2.3 The Delta Method for EV Impact Assessments

This subchapter draws up an alternative approach for assessing the impact of EVs: It is referred to as the Delta Method [3], as it evaluates all arising changes and differences concerning power generation as a direct result of EVs. A reference scenario, which does not consider EVs, provides the basis for this comparison. Hereby all deviations from the reference scenario concerning CO_2 emissions as well as power generation and infrastructure costs are accounted for and allocated solely to EVs.

The assumption that EV integration has no direct effects on the development of power plant infrastructure is merely correct for a small number of EVs and a short period of time, i.e. a few years. The integration of an entire EV fleet of several million vehicles, however, has a noticeable long term effect on the investment and operation strategy of power plant portfolios.

Hence, in order to assess which power plants and their corresponding fuel types make up this delta compared to the reference scenario, power plant simulations with and without EVs are necessary (see Figure 5). This differentiated simulation methodology addresses the differences in the long term development of the power plant infrastructure as well as the changes in their operation patterns.

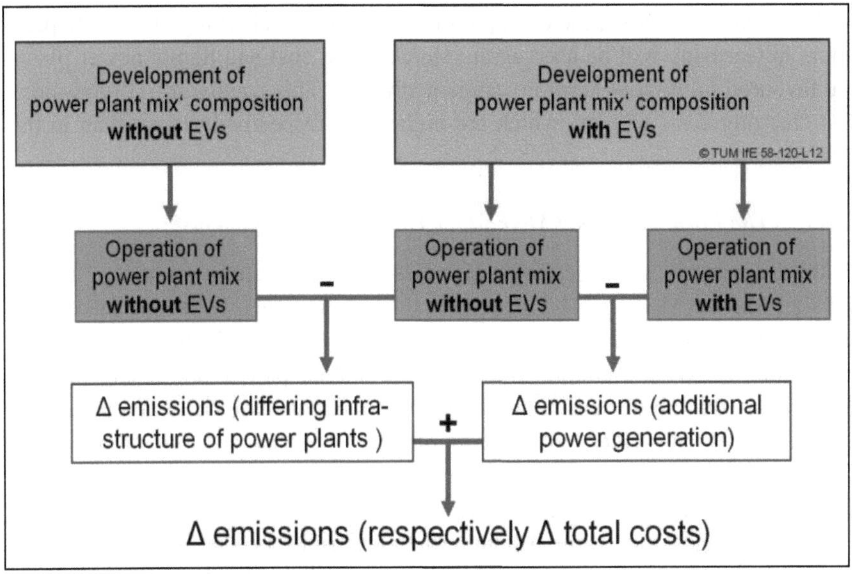

Figure 5: Methodology for allocating and evaluating EV emissions and costs: the Delta Method

Three different scenario simulations are necessary to produce the required deviations concerning power plant infrastructure (blue fields) and operating conditions, which relate to fuel types and their associated emissions and costs (purple fields). The reference scenario, which considers power plant infrastructure and operation without EVs, is depicted on the left side in the flow chart of Figure 5.

The additional electricity can only be generated by power plants, which are either idle or are not already operating under full load. According to the Delta Method these are then the power plants responsible for emissions and costs due to EVs drawing power from the grid. The delta in emissions/costs has two reasons: One is the result of a different power plant infrastructure development over many years, which in general favours power plants with higher investment but lower operating costs, due to their extended full load hours. The second reason is the actual charging demand of EVs. Hence certain power plants will produce the additionally required electricity for EVs. Power plants affected by this, are those ranked in the Merit Order just before the load forecast is covered (as depicted in Figure 3). In a comparison to the reference scenario their full load hours increase and with that all their attributed emissions and costs are allocated to EVs.

To explain the above, the example of controlled EV charging during off-peak hours in Germany will be used again: Hereby hard coal and lignite power plants are favoured in the long term investment strategy. This results in CO_2 emissions and charging costs of EVs, which are higher and respectively lower than in the Mix Method.

3.2.4 *The Parallel Market Method for EV Impact Assessments*

A third approach for allocating EV emissions and costs is by means of the Parallel Market Method [4]. Effectively this method focuses on the manner of power generation according to certain criteria for which electricity consumers are willing to pay a premium, thus creating another parallel electricity market. Specifically this can for example imply a higher quality electricity production for ecological reasons. Hence parallel to offering the basic service of providing power, electric utilities can also market the manner of power generation according to their ecological footprint.

Generally speaking, in the instance when electricity is taken from the grid by the consumer, it was not partitioned and colour-coded beforehand. Although for marketing purposes some organisations prefer to classify and sell green electricity to the consumer, electricity as such is, however, colourless, odourless and not pegged to any ecological trademark. It results from the presence and flow of an electric charge – a result of a process of energy conversion from various primary energy carriers. Hence the primary energy sources and their manner of conversion to electricity determine the higher quality and justify a parallel market for electricity. In Germany and other countries for example consumers are willing to pay higher market premiums for an environmentally friendly power generation based on renewable energies. This parallel electricity market can now specifically be applied to EVs such that emissions and costs of renewable energies are solely allocated to their power demands.

The Parallel Market Method can be based upon a manifold of criteria and specifications. They for instance relate to the kind of renewable energy source, the impact on the environment during construction and operation and among others the handling of supply and demand. Especially the last mentioned criterion offers two pronounced methods of matching supply and demand:

Firstly renewable power supply is completely decoupled from the consumers' electricity demand. Only after a specified period of time (usually after a full year) the amount of renewable electricity generated is required to match the exact amount of electricity drawn from the grid by a consumer. This method is often referred to as “balance-sheeting supply and demand”. TÜV SÜD for example provides the following certificate for such a parallel electricity market: “Certification Criteria: EE01 – Certification of electricity tariffs from renewable energy sources with at least 25 % recently built generation capacities” [5].

The second method of matching supply and demand focuses on synchronising renewable feed-in with the simultaneous power demand of a consumer. It is called “synchronised supply and demand”. TÜV SÜD also provides a certificate for this: “Certification Criteria: EE02 – Certification of electricity from renewable energy sources with a simultaneous supply” [6].

As every other consumer of electricity EVs may also take advantage of this Parallel Market Method by certifying the manner of a higher quality in power generation for their demands and thus justifying their claim as true zero emission vehicles.

An application of the Parallel Market Method with synchronised supply and demand for EVs is shown in the system setup in Figure 6: Depicted is the direct supply through local integration of renewable energy sources into the complex of a single-family house with an EV such as the MUTE.

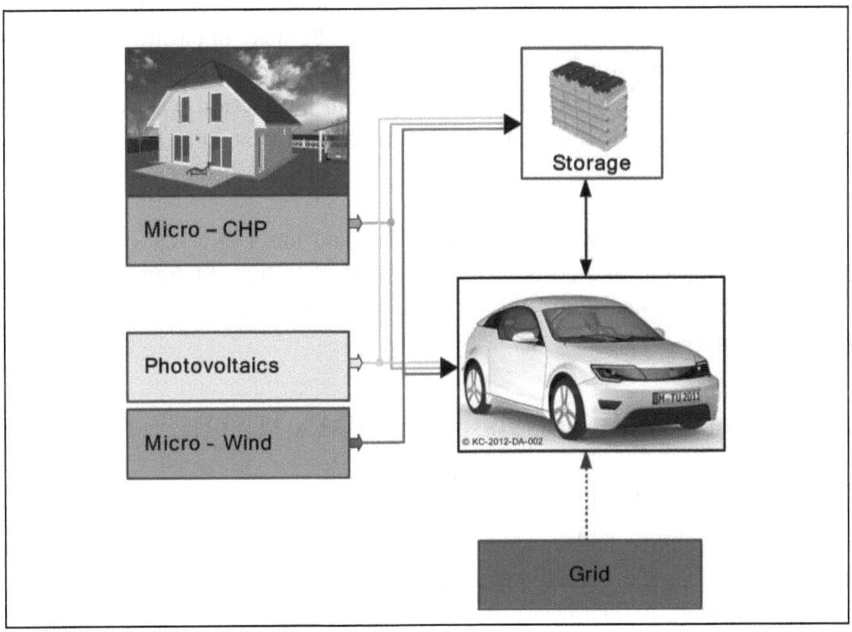

Figure 6: Renewable decentralised charging of an EV in a home environment [7]

Therefore, adapted to the framework conditions of the specific house (building surface area, roof area, location, other limitations), a cost optimization of the needed house energy infrastructure is done to cover the user defined electrical driving needs and as a consequence identify the most cost-effective solution for the MUTE and house owner. The independently size-adjustable components of the system aggregation include both energy conversion devices like photovoltaics and micro-wind turbines and also stationary storage systems like lithium-ion or lead batteries. A heat-led block-type thermal power station (micro-CHP) could also be part of an overall calculation if it had already existed in the house complex before the analysis. Due to high investment costs and as a consequence of uneconomical framework conditions it has been excluded from size optimization. The linear optimization itself is performed by different adapted algorithms like Downhill-Simplex or evolutionary strategy which are optimizing both sizes of the different components and the applied house energy operation strategy.

In order to guarantee secured charging of the MUTE – even in periods of renewable stagnation – and prevent large renewable overcapacities, grid connection is established and electricity purchase in the context of a strict charging criterion is granted. It implies that a maximum 20 % share of electricity purchase from the grid of the hourly charging demand is allowed in 99 % of all hours, at least 80% of the hourly charging demand has to be covered by house internal renewable sources. All cost items are valued using the annuity method using a rate of 5% over a depreciation period of 15 years and are disclosed to the investment date (starting point of the analysis) [7].

The advantages of the Parallel Market Method are a clear cut approach for allocating emissions and costs; hereby primarily EVs benefit from the new renewable power system and only in a second instance other consumers profit from abundances of such a system.

3.3 Simulation Results Based on TUM's MUTE

3.3.1 Simulation Parameters for Modelling the MUTE

In order to quantify and analyse attributed emissions and charging costs of subcompact EVs, the MUTE's electricity demand from the grid is taken as an exemplary value: A final energy demand (Plug2Wheel) of 7,1 kWh/100km can be expected. This considers the effective energy demand for driving purposes under driving conditions of the NEDC (6,6 kWh/100km) [8], losses due to the inverter of the charger (max. charging power of 3,7 kW AC to DC) and losses resulting from electrochemical inefficiencies in the battery (together approx. 7 %). The MUTE's Plug2Wheel demand provides the basis for all simulation results, when applying the three above mentioned EV charging methodologies.

3.3.2 Mix Method: Charging TUM's MUTE with the German Grid Mix of 2015

When using the overall German Power Plant Mix as a basis for calculating emissions and charging costs, the MUTE is considered and defined as a regular consumer of electricity. The vehicle makes no use of centrally controlled intelligent grid management services or decentralised smart meters in order to minimise costs and/or emissions.

In Figure 7 (left column) the share of various power plant types for grid mix charging for the simulation year of 2015 is shown. In comparison to today's grid mix, renewable energies will be taking up a bigger share of the generation mix, while at the same time reducing the demand for thermal power production. The overall emission factor can be calculated to approx. 523 gCO_2/kWh. Considering

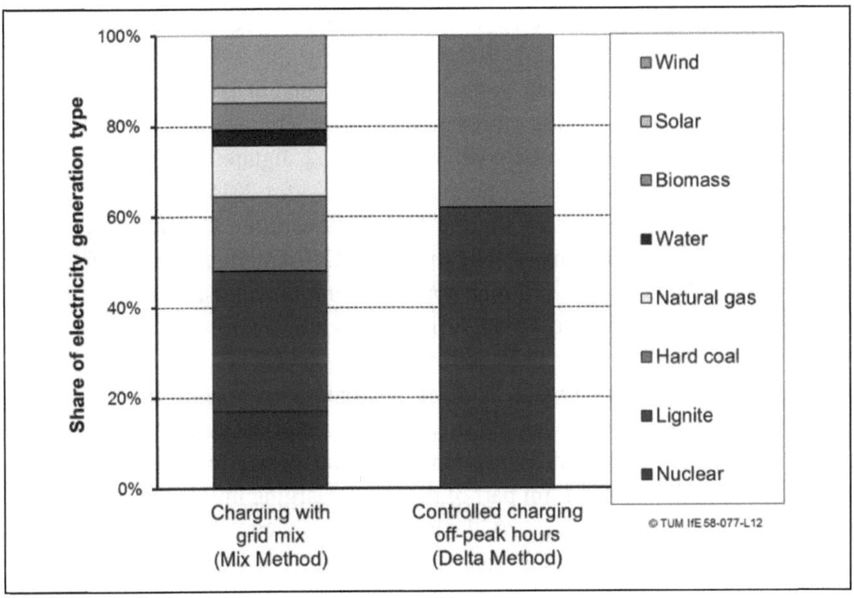

Figure 7: Share of power plants for EV charging according to the Mix and Delta Methods for Germany in 2015

the MUTE's final energy demand, approx. 37 g CO_2/km are thus responsible for driving purposes. Charging costs for electricity would make up about 2,00 €/100 km. This amount already includes an estimate of about 50 % taxes and surcharges as well as additional costs for subsidising renewable energies ("Erneuerbare Energien Gesetz – EEG") predicted for 2015 for household electricity use in Germany.

3.3.3 Delta Method: Cost-Efficient Charging of the MUTE During Off-Peak Hours in Germany for 2015

Figure 7 (right column) displays the share of various power plant types for off-peak charging for the simulation year of 2015. During the early hours of the morning mainly lignite and hard coal power plants will be in a state of partial load or idle. The additional electricity demand for EVs will thus be provided by them ramping up their power generation capacities.

Charging costs and attributable emissions for the MUTE hereby amount to approx. 1,85 €/100 km and 71 g CO_2/km. As explained before, this is a direct consequence of hard coal and lignite fired power plants, which are the next most cost-efficient power plants in the merit order able to provide the additional electricity required by EVs connected to the grid. The high specific emission factors of lignite (400 g CO_2/kWh_{therm}) and hard coal (335 g CO_2/kWh_{therm}) thus account for higher CO_2 emissions than compared to the general power plant grid mix (Figure 7, left column). On the other hand, however, charging costs are lesser, as a result of low fuel costs (hard coal: 9 €/MWh_{therm}, lignite: 4 €/MWh_{therm}) and prices for CO_2 emission credits of predicted 22 €/t_{CO2} for 2015.

In other countries EVs such as the MUTE have quite a different environmental impact than in Germany. The diverging CO_2 emission results of EVs in Figure 8 reflect the vastly different power generation infrastructures of a few exemplary countries selected for analysis: Due to the composition of the French, Swiss and Norwegian power plant infrastructures, the MUTE would only be responsible for about 2 g CO_2/km. This is predominantly due to nuclear power plants in France, hydro power plants in Norway and an almost equal mix of nuclear and hydro power in Switzerland. Thermal power plants burning fossil fuels at very seldom times form part of the EVs' charging mix in these countries, but may be neglected. Countries with similar power plant infrastructures such as the three discussed, can certainly state to have true zero emission vehicles.

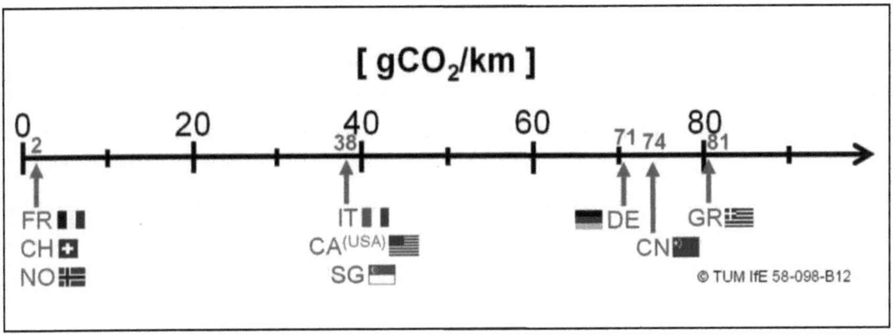

Figure 8: EV emissions scale: specific CO_2 emissions attributable to the MUTE in various countries for 2015 (Delta Method)

In countries with a power generation infrastructure based mainly on natural gas, such as in Italy, California (USA) and Singapore, EVs such as the MUTE are responsible for approx. 38 gCO_2/km. Low CO_2 emissions result from high gas turbine and combined cycle power plant efficiencies as well as from a low specific emission factor of natural gas (202 gCO_2/kWh_{therm}). In China hard coal fired power plants are used to cover the middle load during off-peak charging hours of EVs. CO_2 emissions of about 74 gCO_2/km are hereby attributable to these EVs. In comparison to Germany emissions are slightly higher due to less efficient coal power plant technologies in operation. In countries such as Greece with a predominantly lignite powered infrastructure approx. 81 gCO_2/km can be considered as a direct result of using MUTE-like EVs. This is mainly due to the specific emission factor of lignite, which is the highest among all fossil fuels.

Comparable vehicles with a similar gross weight but with an efficient internal combustion engine have Tank2Wheel consumptions of about 25 kWh/100km for the NEDC. This number is calculated from the same effective energy demand of conventional and electric vehicles with similar driving resistances as a result of sharing all components except for the drive train (ICE and fuel tank versus electric drive train with battery). In the automobile industry the term "conversion design" often describes this method of platform sharing between certain vehicle types. It provides the best basis for comparable calculations on efficiencies and energy consumption. Therefore the CO_2 emissions of a similar vehicle such as the MUTE but with ICE amount to approx. 67 gCO_2/km, considering the specific CO_2 emissions of diesel fuel of 266 gCO_2/kWh.

Taking the above into account the following general conclusions can be drawn: In countries with a similar lignite power generation infrastructure such as

Greece, EVs would unfortunately worsen the country's carbon footprint considerably. For countries such as Germany and China with hard coal power plants providing the middle load, a shift to e-mobility has no immediate environmental benefit. Countries with a strong gas-powered generation infrastructure such as Italy, California and Singapore profit from a noticeable reduction. While countries such as France, Switzerland and Norway (nuclear and hydro powered generation infrastructure), emerge as the major environmental beneficiaries effectively capable of strongly reducing the emissions of the automobile sector in the short term.

Yet in summary all considered countries except Norway would be dependent on thermal power plants to provide the additional electricity needed during off-peak charging hours of EVs. This results in either high or moderate CO_2 yields per km or additional nuclear waste accumulating due to a switch from conventional vehicles to EVs. Hence the question arises, is there an immediate, more sustainable alternative?

3.3.4 Parallel Market Method: Charging the MUTE with Renewables in a Scenario with German State Incentives

The suburban location, which the simulation is based on by means of specific weather data, is characterized by good photovoltaics (PV) and rather average wind suitability. The single-family house includes a building area of 120 m², a roof area of 60 m² and an individual garage with an appropriate charging facility. The MUTE is therefore charged using a standardized household socket (maximum charging power 3,7 kW) controlled by load management to get an equalized charging both daily at off-peak periods and during night hours.

Both investment and electricity costs and feed-in tariffs for 2015 are extrapolated from historical values and actual political regulations. Lifetimes of the energy conversion devices (wind, PV) are both set to 20 years, lifetime for electrical lithium-ion-storage to 2.500 (~10-15 years) full cycles.

Under the assumption of a typical driving purpose as a second vehicle – including an annual distance of 15.600 km and a depreciation period of the renewable energy systems of 15 years – new installation capacities of 9 kW PV (needed roof area 60 m² corresponding 100 %), 2 kW wind power and 2 kWh electrical storage (lithium-ion storage) result as a consequence of an hourly based simulation.

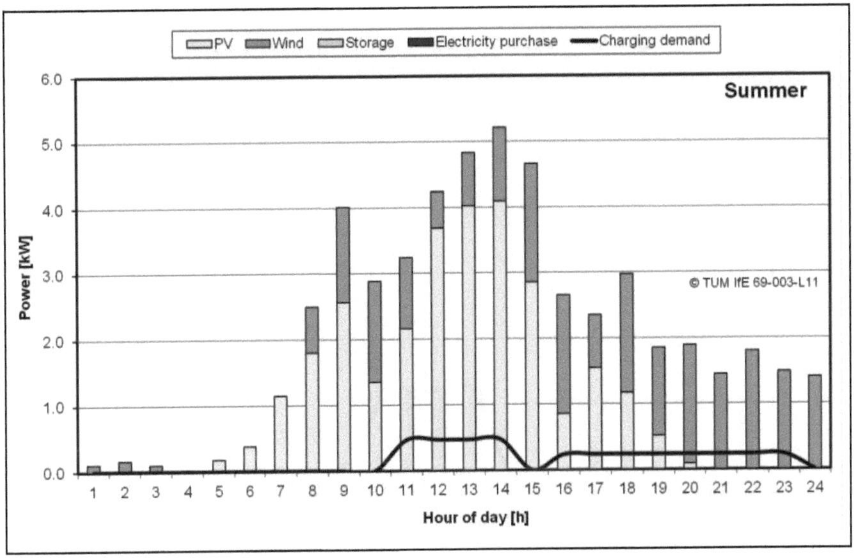

Figure 9: Exemplary daily coverage of the MUTE's charging demand in summer

Figure 9 illustrates the needed high overproduction of the installed overcapacities and furthermore high feed-in tariffs due to the fact of the strict renewable charging criterion (renewable share of the charging demand of 80 % in 99 % of hours a year). As shown, in summer most of the time there is no need for using the electrical storage.

On the other hand Figure 10 shows the exemplary coverage of the charging demand during an extreme winter day. This figure also shows the behaviour of the total system at periods of renewable stagnation including high usage of storage and electricity purchase.

Figure 11 displays the monthly comparison of renewable electricity production and charging demand on the balance sheet including the already mentioned overproduction of renewable energy due to the strict renewable charging criterion.

By maintaining the needed criteria of the simulation, annual driving costs of about 50 € result. On the one hand, these costs consist of annual expenses due to infrastructure, maintenance and operation of the renewable energy systems and on the other hand of electricity purchases and renewable feed-in tariffs. On closer examination of the total costs the large share of renewable tariffs for solar

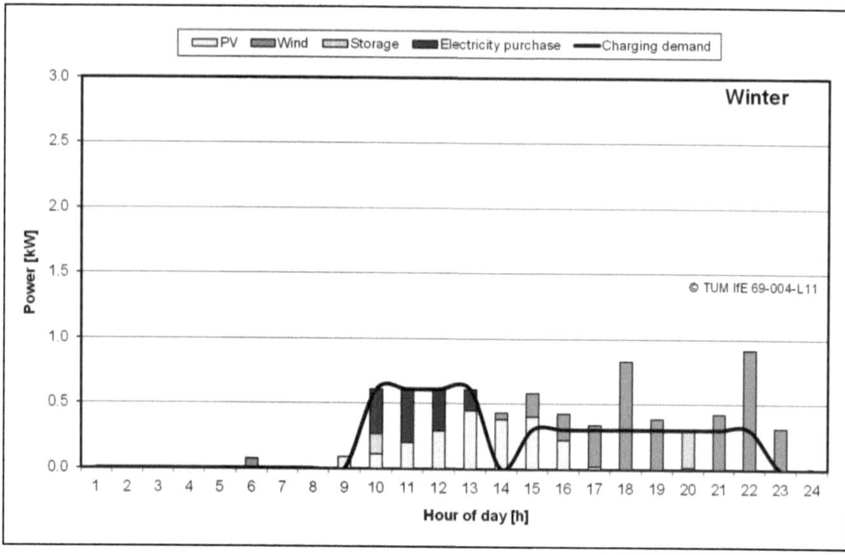

Figure 10: Exemplary daily coverage of the MUTE's charging demand in winter

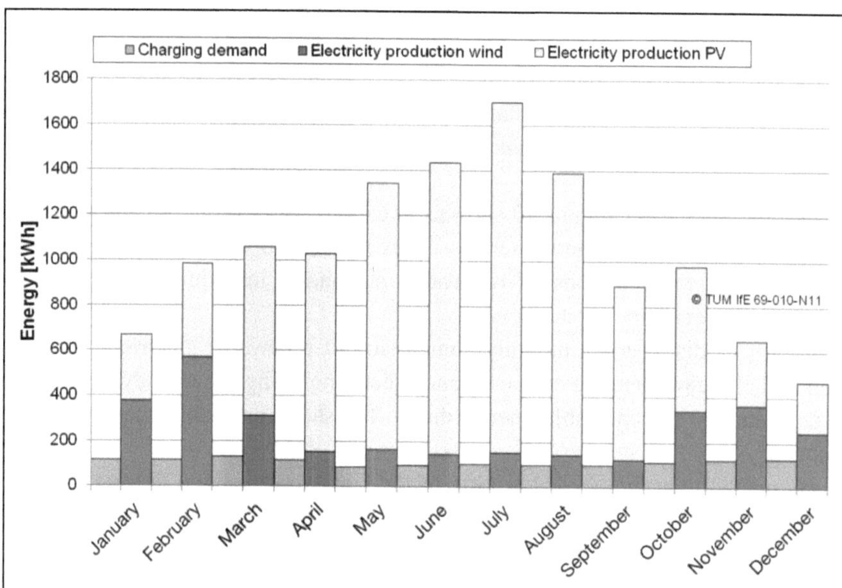

Figure 11: Comparison of Renewable Electricity Production and Charging Demand on the Balance Sheet

produced and either own consumed or fed-in electricity is clearly recognizable. Over a total observation period of 15 years, nearly a compensation of the infrastructural costs (~ 23.000 €) by the renewable feed-in tariffs takes place. The electricity purchase from the grid – optimized by the applied operation strategy – is limited to 13 kWh a year (~ 1 %), which has the effect of a very small share of electricity purchasing costs and also of the status of a zero emission charging system (0,4 gCO_2/km). At a direct apportionment of the infrastructural and operational costs to the total renewable produced electricity, direct power generation costs of about 0,12 €/kWh result [7].

3.3.5 *Sustainable Mobility: Summary of the MUTE's Ecological and Economical Competitiveness to ICE's*

As mentioned above, charging costs and attributable emissions of the MUTE strongly depend on the primary energy chain. At the Mix Method approx. 2,00 €/100 km and 37 g CO_2/km, at the Delta Method 1,85 €/100 km and 71 g CO_2/km and at the Parallel Market Method 0,30 €/100km and 0,4 g CO_2/km result. In comparison to a similarly powered and sized internal combustion engine vehicle (ICV) (diesel consumption: 3,3 l/100 km) [9] and a predicted average diesel price for 2015 of 1,49 €/l [10], all charging scenarios show lower driving costs and emissions (also see Table 1).

Table 1: Comparison of driving costs and CO_2 emissions of ICV and different methods of supplying MUTE

 Source: Daimler	 Source: FTM		
Daimler Smart	MUTE		
ICE Diesel	**Mix** Method	**Delta** Method	**Parallel Market** Method
86 g CO_2/km 4,92 €/100km	37 g CO_2/km 2,00 €/100km	71 g CO_2/km 1,85 €/100km	0,4 g CO_2/km 0,30 €/100km

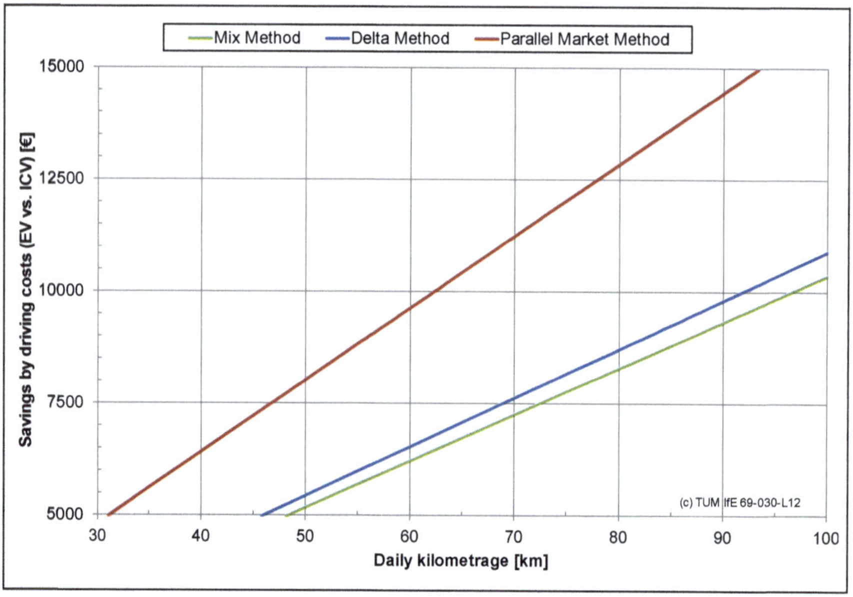

Figure 12: Savings by driving costs of EVs in contrast to conventional ICVs compared by different supply methods (assumed vehicle lifetime 15 years).

As depicted in Figure 12, a daily kilometrage of only approx. 31 km (Parallel Market Method), 46 km (Delta Method) or 48 km (Mix Method) are required for cost savings of about 5.000 € over an assumed vehicle usage period of 15 years. Under the assumption that investment costs of the MUTE should be as high as investment costs for conventional ICV and in addition monthly costs for battery leasing of about 50 € [11] exist, a kilometrage of approx. 56 km/day (Parallel Market Method), 82 km/day (Delta Method) or 87 km/day (Mix Method) are needed to achieve cost parity between MUTE and ICV (rate = 7%, depreciation period = 15 years).

3.4 Conclusion

The introduction of electric vehicles such as TUM's MUTE into the mass market over the next years poses the question, where EVs will find the most favourable conditions concerning economical and environmental criteria. There is no doubt that low operating costs and environmental implications due to EVs provide the best basis for moving out of the niche and into the mass market. Hence the crucial factor for a market breakthrough of EVs lies with the type of power plants in the upstream primary energy chain, which de facto determine operation costs and attributed emissions of EVs. Countries such as France and Switzerland emerge as major environmental beneficiaries effectively capable of strongly reducing the emissions of the automobile sector in the short term. At the same time the additional electricity can be provided by existing nuclear and/or hydro power plants at very competitive costs.

Yet in summary all considered countries would be dependant mainly on middle load thermal power plants to provide the additional electricity needed during off-peak charging hours of the MUTE. This results in either high or moderate CO_2 yields per km or additional nuclear waste accumulating due to a switch to electric mobility. On a long term basis it will therefore become inevitable for the power generation sector to switch to predominantly renewable sources. The long lasting aftermath of CO_2 emissions and nuclear waste attributable to EVs can hereby be avoided. Coupling the power demand of EVs to the fluctuating feed-in of renewables thus has a high potential to foster sustainability in both the power generation and the automobile sector.

In conclusion our studies show that sustainable mobility on the basis of highly efficient EVs such as TUM's MUTE in combination with renewable energy charging is the cleanest and most affordable option in Germany to date. Additional EV investment incentives such as in France or the USA are therefore not necessary in Germany: With the help of the EEG, low charging costs and thus EV operation costs arise. Hence coupling EVs and renewable energies provides a double benefit for society: clean energy generation combined with next generation mobility.

3.5 References

[1] Bundesministerium für Verkehr, Bau und Stadtentwicklung (BMVBS) "Datensatz Mobilität in Deutschland 2008", DLR Institut für Verkehrsforschung, Berlin 2011.

[2] Kuhn, P.; Kühne, M. "Optimierung des Kraftwerks- und Speicherausbaus mit einem iterativen und hybriden Modell", 9. Fachtagung: Optimierung in der Energiewirtschaft, VDI-Berichte Nr. 2157, VDI Verlag GmbH, Düsseldorf 2011.

[3] Roth, H.; Kuhn, P.; Gohla-Neudecker, B. "Sustainable Mobility – Cost-Effective and Zero Emission Integration of Germany's EV Fleet", Procee-

dings of 2nd International Conference on Clean Electrical Power – Renewable Energy Resources Impact, IEEE, Capri (Italy) 2009, pp. 207-211.

[4] Gohla-Neudecker, B.; Roth, H.; Wagner, U. "Nachhaltige Mobilität: Anwendungsscharfe Bereitstellung von erneuerbarer Energie für Elektroautos", Energie-wirtschaftliche Tagesfragen 60. Jg., Heft 4, etv Energieverlag, Langfassung unter www.et-energie-online.de, Essen 2010, pp. 81-86.

[5] TÜV SÜD Industrie Service GmbH. "Certification Criteria: EE01 – Certification of electricity tariffs from renewable energy sources with at least 25 % recently built generation capacities", File: CMS-Z: E-Cert 80, version 10, Munich 2008.

[6] TÜV SÜD Industrie Service GmbH. "Certification Criteria: EE02 – Certification of electricity from renewable energy sources with a simultaneous supply", File: CMS-Z: E-Cert 82, version 10, Munich 2008.

[7] Kandler, C. "Energiewirtschaftliche Optimierung der hauseigenen erneuerbaren Stromerzeugung zur Versorgung eines Elektrofahrzeugs", Diploma Thesis at the Institute for Energy Economy and Application Technology, TU München 2010.

[8] Lehrstuhl für Fahrzeugtechnik (FTM) of Technische Universität München „NEDC simulations of final energy demand of the MUTE", Munich 2011.

[9] Daimler Smart Fortwo, http://www.smart.de, 2012.

[10] Bundesministerium für Wirtschaft und Technologie (BMWi) "Die Entwicklung der Energiemärkte bis 2030 – Energieprognose 2009", IER, RWI, ZEW 2010.

[11] Renault ZE Twizy, http://www.renault-ze.com, 2012.

III Powertrain

4 Highly Integrated Axial Flux Drive for Low-Voltage Applications

Claus Mueller, Bernhard Hoffmann

Abstract: A coming market segment of lightweight vehicles for urban mobility is bringing further challenges in terms of low weight and cost effectiveness. On system level, the usage of a low voltage system is desirable. A transverse flux permanent magnet motor design seems to be very well suited to provide high torque and power over a wide speed range at low weight, volume and production cost. A corresponding drive unit with integrated inverter has been developed for automotive application. The motors design and its operational behaviour will be discussed. A short description of the cost-effective manufacturing process is given.

Keywords: Compact, lightweight, low-voltage, integrated automotive drive system

4.1 Introduction

The market penetration of electric drives within HEV's and BEV's is mainly challenged by massy and expensive batteries. As a consequence, a quick market penetration of these vehicles is impeded by an increasing curb weight in conjunction with low operating range and high prices. The currently available models are therefore not suited for volume sales. A new class of lightweight electric vehicles targeted for individual urban mobility seems to have the potential to speed up the market penetration of BEVs.

Scooters, ATVs and Micro compact cars do not have a high renown due to a usually cheap make in conjunction a small and noisy engine. This situation can be fundamentally changed by adding premium properties in combination with a smooth and silent electric drive into this class of vehicles. The result will be beneficial in several dimensions and can help to boost the spreading of electric vehicles:

- The lowered weight is significantly reducing the total energy consumption and consequently lowering the batteries capacity requirements.
- Specifically targeted for urban mobility, the operating range which is expected by the customers can be easily handled.
- Low weight and a limited top speed are reducing the motor power requirements.
- A low-power drive can be implemented using a cost-efficient low-voltage system.

Overall: The surplus costs imposed by the vehicle electrification are comparatively small. They are accompanied by a completely new customer value for urban mobility.

4.2 Low-voltage system

Low voltage systems for vehicle traction motors have numerous advantages. For European markets, ECE-R 100 is defining nominal voltages larger than 60 V DC as automotive high-voltage system. It is also establishing basic requirements for protection measures. For low-voltage systems ≤60 V no comparable regulations apply.

4.2.1 On-board power supply

A HV battery requires terminals with ingression protection measures and a separate precharging connector is necessary. For LV batteries the BMS means less effort and cost due to a much smaller number of cell stack levels.

For HV systems, an insulated bipolar wiring is required. The used cables usually have a double insulation with screen and special care has to be taken regarding the connectors. Galvanic insulation barriers are required for power and signal lines within HV components. The wiring of LV-systems can be done unipolar similar to automotive 12 V supply networks. Fuses rated for 60 V application are sufficient. The vehicle chassis may serve as screen for radiated emissions.

4.2.2 Power electronics

In the future, the common 12 V on-board power supply will remain independently of a second higher voltage system. For continuous feeding of the 12 V supply a DC/DC converter is required. For LV application the converter does not need any galvanic insulation and can be designed for peak efficiencies of up to 97 %. MOSFET semiconductors can be used and air cooling is usually sufficient.

Regarding the LV-inverter, very fast switching MOSFETs can be used instead if IGBTs for HV inverters. Thanks to the MOSFETs body diode, no separate freewheeling diodes are required. A higher level of system integration is possible, because clearances and creepage paths are significantly smaller. For the DC link capacitors the use of cheap aluminium electrolytic capacitors is advantageous, because a low voltage rating on the other hand is leading to higher temperature ratings (>105°C) or higher life times (>10.000 h) as they are required for automotive applications.

For assembly, maintenance and assistance trained personnel is required for manipulations regarding the HV components.

The sum of all these issues enables a significant potential for cost savings, if a low-voltage system is employed.

We consider single units up to 25 kW of shaft power as very well suited for low voltage supply. Two of these units can be mounted to drive a vehicles axle without the need for a differential gear. A total power of 50 kW is adequate for a vehicle curb weight up to 800 kg.

Motor Design and Data

A three-phase single-sided permanent-magnet claw-pole transverse-flux machine design with an outer rotor was chosen as a favourable design for high shaft power at low voltages in combination with compactness, low weight and cost-effective manufacturing.

Due to a significant axial component of the stator flux, we also refer to this design as an axial flux design.

4.2.3 Motor Design

The claw-pole stator design allows an independent design of the magnetic and the electric circuits. There are no conflicting goals of allocating space for stator teeth and for the winding, as they are typical for radial flux machines. Furthermore the stator winding does not have any winding heads and consequently no corresponding copper losses. Thus high cross-sections of copper are allowable and copper losses are on a quite low level.

A high number of pole-pairs is generating a high level of thrust inside the air-gap.

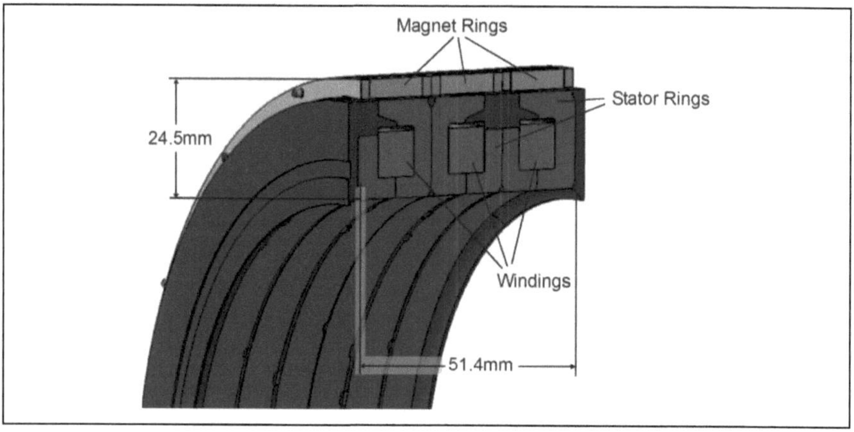

Figure 1: Active part of the motor

In order to provide maximum torque with a given outer diameter, an outer rotor design is the optimum solution. Because the rotor has a much smaller radial dimension compared to the stator, the air-gap diameter can be maximised. First, the active air-gap area providing thrust is maximized, second the thrust's lever is also maximized resulting in a maximum volume-specific torque.

The no-compromise machine design on the other hand imposes some extra-requirements for the inverter. Due to the high stray inductivity, a low power factor and a higher reactive power is demand of the inverter. On the other hand, the high stay inductivity allows field weakening over a very wide speed range. This allows the inverter to operate at limited voltage even at high speeds. The rated inverter power remains low.

Top speed is about four times the nominal velocity. A high number of pole-pairs in conjunction with a high rotor speed lead to high fundamental frequencies, which have to be handled by the inverter.

The total active volume for generation of up to 75 Nm of torque and 20 kW continuous power is a ring-shaped body consisting of six identical stator rings encompassing the stator windings and three magnet rings on the rotor side. These parts require a volume of just 25.1 cm^3 and weigh about 4 kg.

Essons' power density factor C defined by $M = C * d^2 * l$ is reaching a very good 5.3 $kWmin/m^3$.

4.2.4 *Stator Design*

Claw pole design is known from rotors with electric excitation of synchronous machines. A typical example is the 3-phase alternator of a passenger car. However, the claw poles of the DYNAX® motor are utilized to conduct the stator flux. The principle of the stator construction and the path of the magnetic field lines are depicted in Figure 2. By the claw pole segments, the magnetic field lines are conducted closely around the litz wire winding. They mostly leave the claw poles in radial direction into the motor's air gap, and close tangentially above [2], [3].

In contrast to radial flux machine designs, three-dimensional flux conduction has to be managed. Accordingly a typical Fe-Si lamination which is capable just of a 2-dimensional flux direction cannot be utilized. Bonded iron powder materials are necessary to avoid eddy currents respectively iron losses.

Recent developments provide magnetic materials comparable to state of the art 0.2 mm laminated sheet stacks regarding induction, iron loss and heat resistance. These materials are improved SMCs (Soft Magnetic Composites), for example MBS318 from PMG.

Six identical stator rings are manufactured in a high-pressure, high-temperature sintering process. The sintered rings receive a heat treatment and do not need any further machining before assembly. This is compensating high costs for the tooling in a volume manufacturing process. A circular water cooler unit is also serving as stator carrier. The stator rings are soldered to the water cooler for optimum heat transfer. The stator is completed by integration of a position sensor and a resin potting process for an enduring tolerance of harsh environmental conditions.

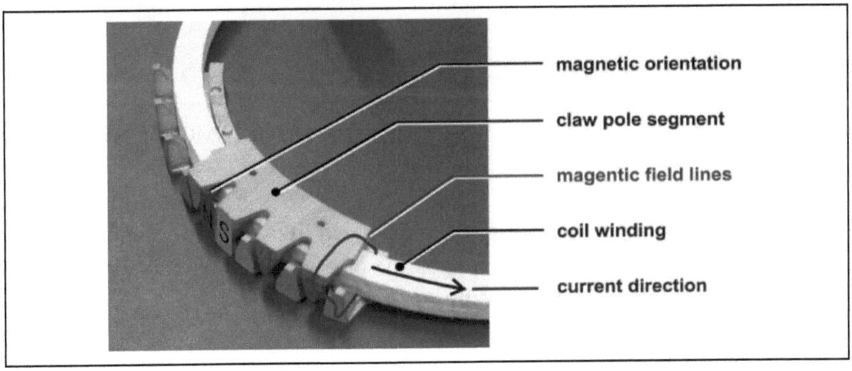

Figure 2: Claw pole segments and winding

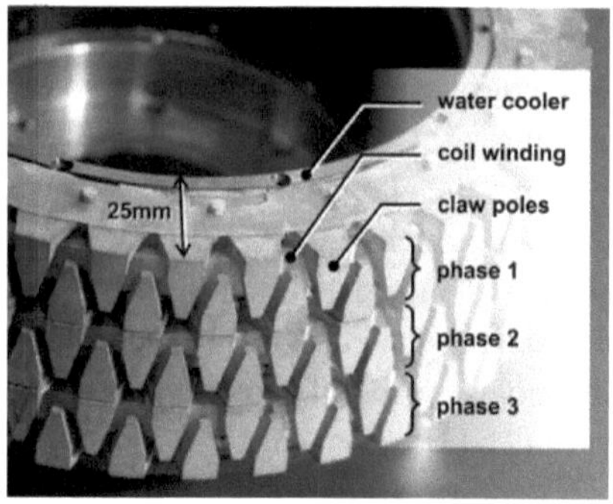

Figure 3: Finished stator with resin potting

Figure 4: Claw pole stator assembly

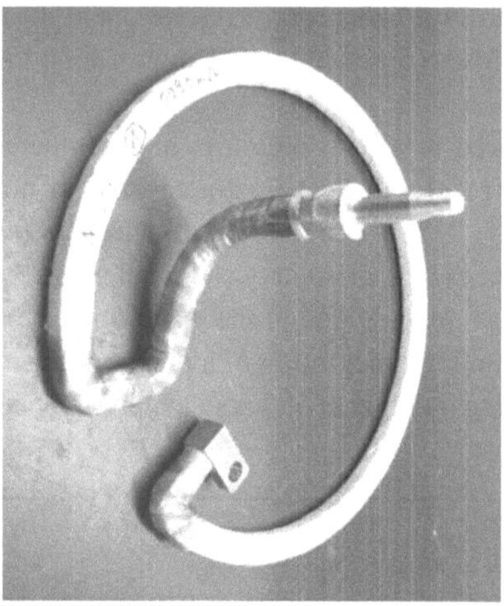

Figure 5: Single stator winding

Figure 6: Sintered SMC stator ring

The stator winding is made of pressed insulated litz wire of rectangular cross-section. Two copper terminals, one for the phase terminal, one for the star-connection are soldered to the winding.

4.2.5 Rotor Design

The bell shaped rotor is an outer rotor with a thin layer of permanent magnets applied at its inside. It is made of glass fiber reinforced plastic (GRP). A back iron is not applied. The construction of the rotor is depicted in Figures 7 and 8.

The orientation of the permanent magnets corresponds to the magnetic orientation of the claw poles (see Figure 2), as the winding is fed with current. The magnet rings have a 120-degree shift between each phase. Inside the rotor bell, a slotted sensor disc for the polar angle detection system is located. The system is based on an inductive distance measurement. It detects the notches of the aluminum disc and generates a sin/cos signal.

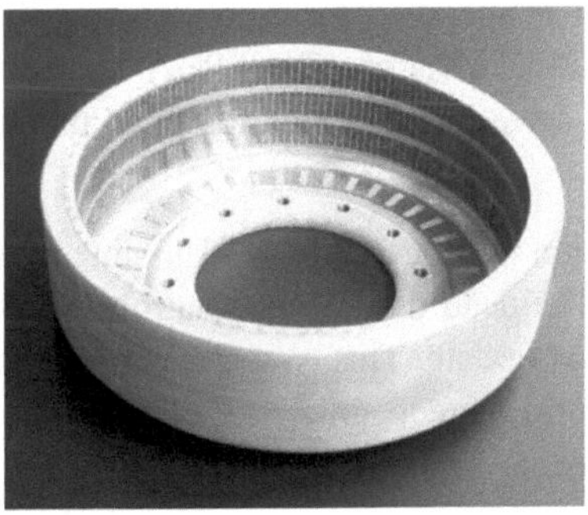

Figure 7: Finished bell shaped rotor

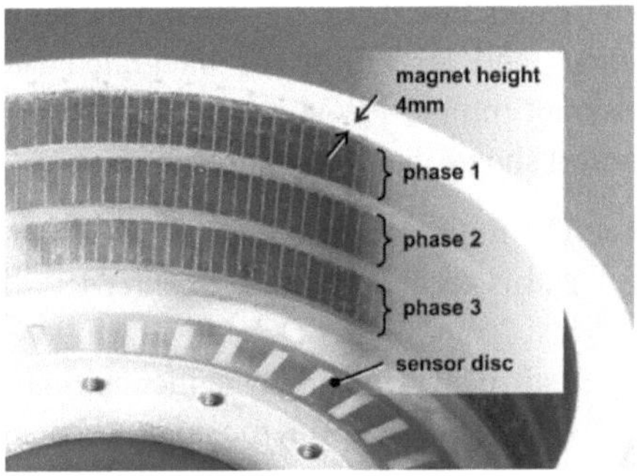

Figure 8: Magnet rings of the rotor

4.3 Control and Operational Behaviour

4.3.1 Characteristics of the Motor

In comparison to conventional PMSM-motors, the DYNAX® motor has some specific characteristics:

- The current for maximum torque demand $I_{S,max}$ is the continuous current.
- The motor requires high fundamental frequencies of the feeding system: 2p = 40 means fp = 4 kHz at 6000 rpm.
- The voltage drop of the stray inductance can be used for a field weakening over a wide speed range – an important feature for starter-alternator applications.

- The high stray inductance causes a short circuit current I_{SC} which is below the motor's rated current. This is a significant advantage, because the safe failure state at high motor speed is a short-circuiting of the phase terminals using the inverter. The value can be derived according to the following equation:

$$I_{SC} = \frac{E_s[V_{peak} / Hz]}{2\pi * L_S[H]} = \frac{20mV / Hz}{2\pi * 12.2\mu H} = 260A_{peak} <= 450A_{peak} \approx \frac{1}{2} I_{s,\max}$$

4.3.2 Motor Control

The control hardware consists of a 100x55mm-board with microcontroller, FPGA and Ethernet/CAN communication. It is specified for automotive applications and is part of the integrated MOSFET inverter. The FPGA controls interrupt timing, PWM-generation and error detection. The field oriented motor control shows the following features:

- The field oriented control (FOC) algorithm is triggered by a 25 µs-interrupt and is feeding a 25 kHz center-PWM. The dead time of the control loop is just 25 µs from ADC-conversion to PWM-update. This way it is possible to realize a fast and dynamic current control. Steep gradients of rotational velocity (up to 50 Krpm/s) can be handled.

- At field weakening operation, the converter output voltage U_s is used at least prior for driving the torque generating current-component $I_{s,q}$. Here the field weakening current $I_{s,d}$ is driven by the difference of the back-EMF and the remaining q-component of U_s. This current does never exceed the maximum value of the short-circuit current, which is reached at $U_{s,q} = 0$. Due to this effect, the drive's power can be kept constant over a very wide speed range, while the inverter voltage does not have to be increased.

- The rotor angle can also be determined very exactly by a computation of the rotor's EMF and it is possible to realize a sensor less control, running very stable and without any disturbances.

The operating states for a field oriented control at motor operation with maximum torque are demonstrated graphically by the vector diagrams in Figure 9. If 15 V are applied to the DC-link, the inverter can provide a phase voltage U_S of maximum ~ 10 V peak value (the motor is configured in Y-connection). The inverter's operation range is indicated by the dotted circle in the diagrams.

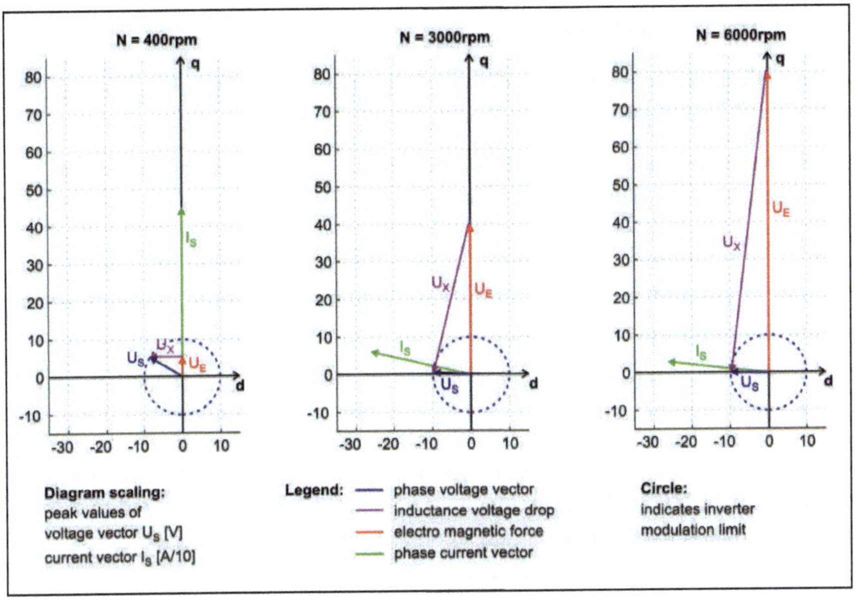

Figure 9: Vector diagrams at VDC = 15V and maximum motor torque / power

When the maximum torque generating current component (in q-direction) is set, the voltage limit is reached at 400 rpm. This can be seen in the left diagram of Figure 9. For higher speeds, operation in field "weakening" mode is necessary. The current vector is rotated to left by an additional field compensating current in negative d-direction, and the voltage drop of the inductance U_X can be utilized to keep U_S inside the 10 V-circle.

The vector diagrams indicate that the motor's EMF can take values which are significantly above the operation range of the inverter. At 6000rpm, inverter and motor control have to manage an U_E which is 8 times above the inverter voltage! As mentioned in section 2.1, the secure state of the system is a short-circuiting by the inverter, which has to be applied if any error occurs or if the motor control is inactive. This is realized by a special, independent operating function implemented in the FPGA, which controls the PWM signals. An additional circuit detects hardware failures and can directly activate the inverter's low side switches to short-circuit the motor terminals. This circuit is supplied directly by the DC-link voltage – this way it can protect the system even if the motor is driven and no external voltage is supplied to the inverter.

4.4 Power Electronics

Motor is designed with an integrated inverter/controller unit (IICU). The integration is beneficial regarding the following aspects:

- Lower total weight and volume because of multiple usages of parts of the case and of coolant connectors and ducts.
- Lower weight, losses and cost by omitting motor cables and connectors.
- Optimized EMC/EMI characteristics due to a single closed containment for inverter and motor.
- Higher reliability and lowered assembly cost due to a reduction of the number of parts in use.

The half bridges are directly connected to the phase terminals of the motor. The effort to fulfil EMC/EMI requirements is significantly lower compared to a solution based on a separate inverter and a connection via motor cables.

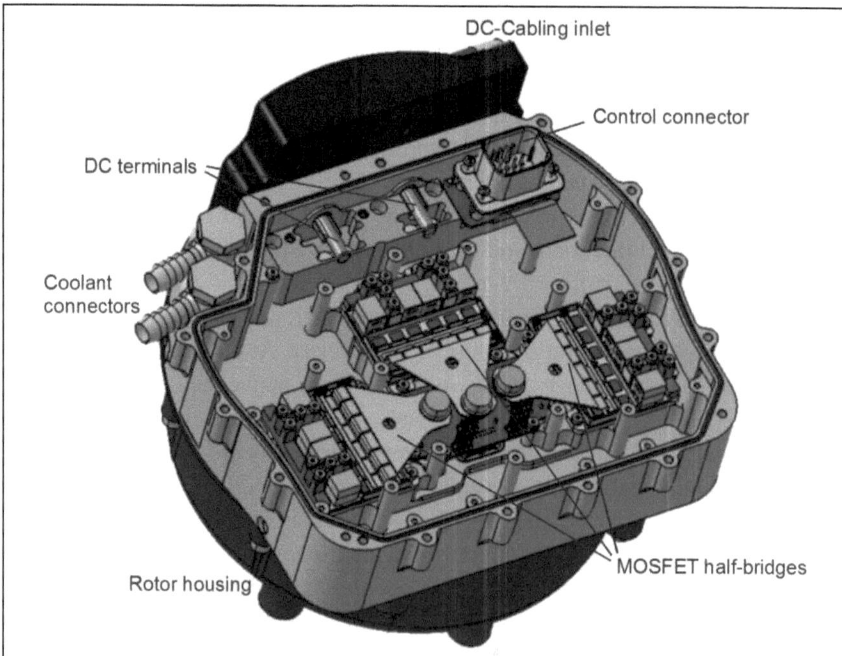

Figure 10: Integrated inverter without DC-link, main controller board and rear cover

The MOSFET half-bridges have been designed for an exceptionally low inductance and incorporate film-type snubber capacitors on the DCB. This enables the usage of a low R_G in conjunction with a very good utilization of the MOSFETs blocking voltage and leads to a significant reduction of the switching losses.

4.5 Measurement Results

According to the diagrams depicted in Figures 11 and 12, the motor is covering a speed range of ~7000 rpm in constant power regime. A very good ratio n_{max}/n_{nom} of 3.5 can be reached. The attainable shaft power is slightly decreasing for high rpm's. This effect is caused by radial rotor bell expansion leading to a widening of the air-gap.

Maximum power is only slightly above the continuous power of 20 kW. A significantly higher peak power can be delivered by the motor, but this would require higher voltages.

The system peak efficiency is comparable to a grid-fed high-voltage 20 kW induction motor. The small unit size, high currents and high pole alternating frequencies are impeding higher values as they are known for larger high-voltage drive systems.

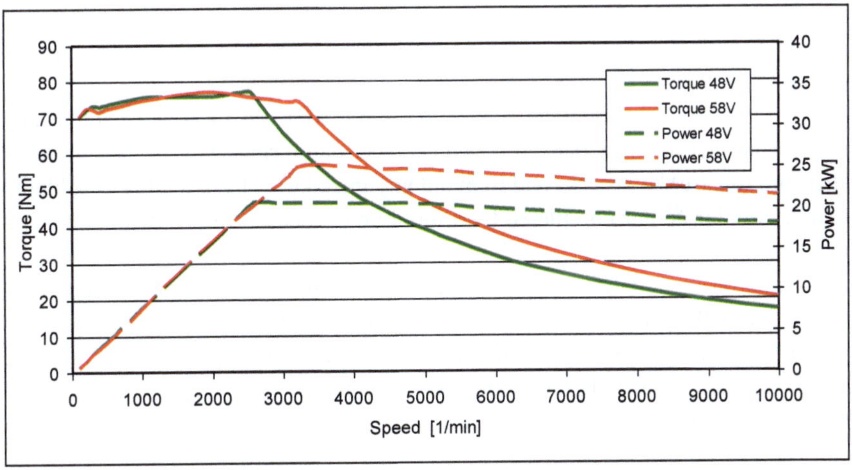

Figure 11: Power characteristics

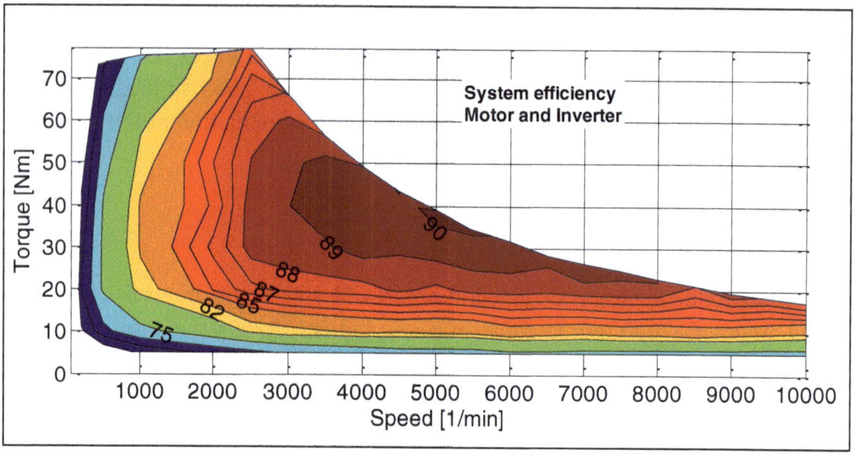

Figure 12: System efficiency diagram

4.6 Assembly

Component manufacturing and motor assembly can be done notably effortless. The process is well-suited for partial or full automation.

The rotor's magnet rings are automatically assembled by a pick and place unit. They are shoved onto a wrapping drum and the glass-fiber tissue is reeled around. As final manufacturing step the resin is injected by a resin transfer mould (RTM) process.

For the stator, six identical power iron rings sintered of SMC material and three windings are just packed together. The stator rings have a self-aligning structure to ease the process and to avoid any mistakes. The assembly is soldered onto a circular cooler completed by a moulding process using a two-component resin.

Compared to radial flux motors either with concentrated or distributed windings, the total manufacturing process has considerably less steps. One could talk just of a line assembly instead of the usual manufacturing process. Because the motor has a low weight and compact dimensions and none of the utilized materials is of a high price, the total production costs are very competitive.

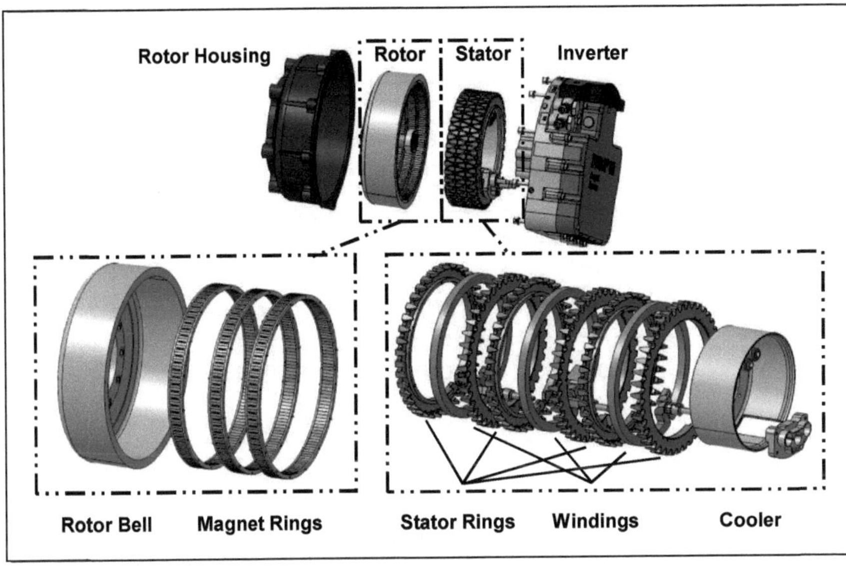

Figure 13: Drive assembly

4.7 Safety Aspects

The DYNAX drive system cannot be only considered safe due to the low nominal voltage and its contribution to the vehicle safety by its low weight, is shows also a very good behavior in case of electrical failures related to the inverter or in case of unintended interruption of the controllers supply voltage.

For high rotor speeds, short circuited terminals are the safe state of the DYNAX motor. The maximum power dissipated at maximum rotor speed is about 1700 W, which is inside the thermally stable boundary of the motor. The short circuit current peaks at about 450 A which can be permanently handled by the inverter. For high speeds the braking torque is below 2 Nm. At low speeds the short-circuit torque has a maximum of 25 Nm, well below the nominal torque.

The open circuit case is safe up to 5000 1/min. At higher speeds, excessive losses and a high rectified back-EMF will damage motor and inverter. On the other hand there is no torque peak at low speeds.

An independent circuitry makes use of this behavior and assures safe and smooth operation in case of electrical failures. The circuitry is autonomously fed

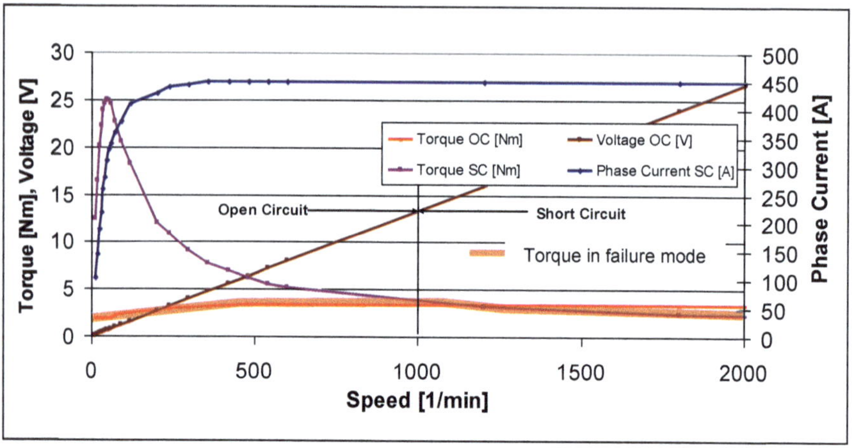

Figure 14: DYNAX failure behavior

by the DC-link voltage and is short circuiting the motor phases by direct driving the low-side semiconductor switches. An almost speed independent torque in failure mode depicted as a bold line in Figure 14 is attained.

This failure mode of the drive system enables the application as a direct-wheel drive even for scooters or motorcycles, because driving stability is hardly affected. The low braking torque in failure mode leads to a smooth deceleration of the vehicle.

4.8 Conclusion

The application of low-voltage claw-pole axial flux motors can provide significant advantages on the system level, e.g. compact dimensions, less protective measures for electric shock, a safe failure mode and low production costs.

Measurement results show a wide speed operation area which is significantly enlarged in comparison to conventional PMSMs.

The presented drive unit represents an ideal solution for powering lightweight electric vehicles and may also be used as a generator unit for range-extenders.

4.9 References

[1] Hoffmann, B.; Kleimaier, A. „Kompakte Elektroantriebe für Hybridfahrzeuge“ 3. Braunschweiger Symposium „Hybridfahrzeuge und Energiemanagement“, Braunschweig, Germany, Feb. 2006.

[2] Gründl, A.; Hoffmann, B.; Rasch, R. „Verbesserte Axialflussmaschine“ Patent application DE102009021703A1, publ. 2009-05-18.

[3] Gründl, A.; Hoffmann, B. ”Permanently excited electrical engine”, Patent application PCT/EP2008/009546, publ.2009-05-22.

[4] Kleimaier, A.; Hoffmann, B.; Scherer, A. “Duplex pulse controlled inverter with a film capacitor DC-link”. 5th Intern. Conf. on Power Electronic Systems (CIPS), Nürnberg, Germany, Mar. 2008

5 Planetary Centrifugal Pendulum Absorber (pCPA) – New type of Centrifugal Pendulum Absorber for Applications in Highly Downsized Hybrid and Range Extender Combustion Engines

Prof. Dr.-Ing. Karsten Stahl, Dr.-Ing. Hermann Pflaum,
Dipl.-Ing. (FH) Georg Johann Meingaßner,
Prof. Dr.-Ing. Dr.-Ing. habil. Heinz Ulbrich, Dipl.-Ing. Johannes Mayer

Abstract: Currently automotive technology is moving more and more from conventional combustion engines to hybrid and fully electrical power drives. The low battery capacity of conventional batteries causes small operating ranges and constrains electrical support in hybrid vehicles.

One possibility to increase the range of electric cars on top of energy recuperation during braking operations is to generate electric energy directly using a generator driven by a range-extender combustion engine. Continuous reduction of emissions and fuel consumption forces the application of supercharged combustion engines with low numbers of cylinders. However, this causes undesirable torsional vibrations even in hybrid and range extender concepts.

The pCPA-concept developed and engineered at the Technische Universität München (TUM) is a new type of pendulum-type absorber suitable for applications in highly downsized and downspeeded conventional powertrains as well as for hybrid and range extender combustion engines. The pCPA enlarges the effective range of pendulum-type absorbers and improves the effectiveness and acoustical performance particularly in transient load states and in operational areas with low rotational speeds.

Keywords: pCPA, planetary centrifugal pendulum absorber, torsional vibrations, automotive powertrain, range extender, combustion engine, plug-in-hybrid vehicle with auxiliary power unit (APU).

5.1 Current Trends in Automotive Technology

Rising fuel prices and tightened emission regulations lead to increased development of more efficient drives, especially in vehicles.

The development trend of fuel- and diesel-powered vehicles is towards low-emission hybrid vehicles and to purely electric driven vehicles with highly downsized and downspeeded combustion engines. [1]

5.1.1 Downsizing and Downspeeding

Main topics in research and development of combustion engines are Downsizing and Downspeeding, targeting high engine power combined with lower fuel consumption and lower emissions.

An outlook for 2020 regarding powertrain technologies (Figure 1) shows that the number of combustion engines will increase due to powertrains including hybrid and range extender combustion engines.

Downsizing results in highly turbocharged combustion engines with low swept volume and less number of cylinders. Engine torque and power is at least equal to conventional engines (Figure 2).

Downspeeding concepts aim shifting engine operating ranges to low speeds and higher loads increasing the efficiency of combustion engines.

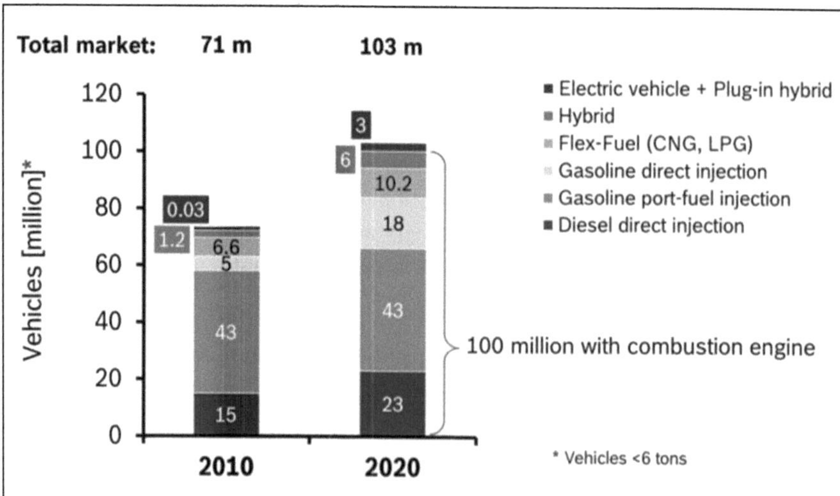

Figure 1: Outlook for 2020 regarding powertrain technologies [2]

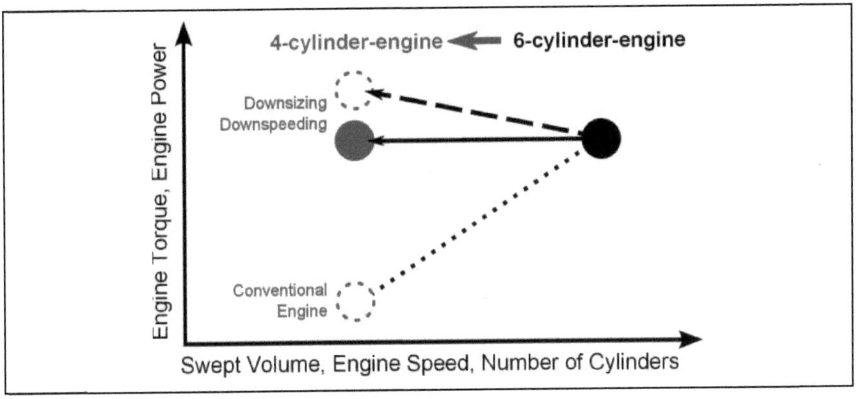

Figure 2: Downsizing and Downspeeding [3, modified]

Downsizing and Downspeeding reduce fuel consumption and pollutant emissions by increasing engine efficiency significantly. Hence, Downsizing and Downspeeding will be a key-factor of the CO_2-reduction targets of the European Union (Figure 3).

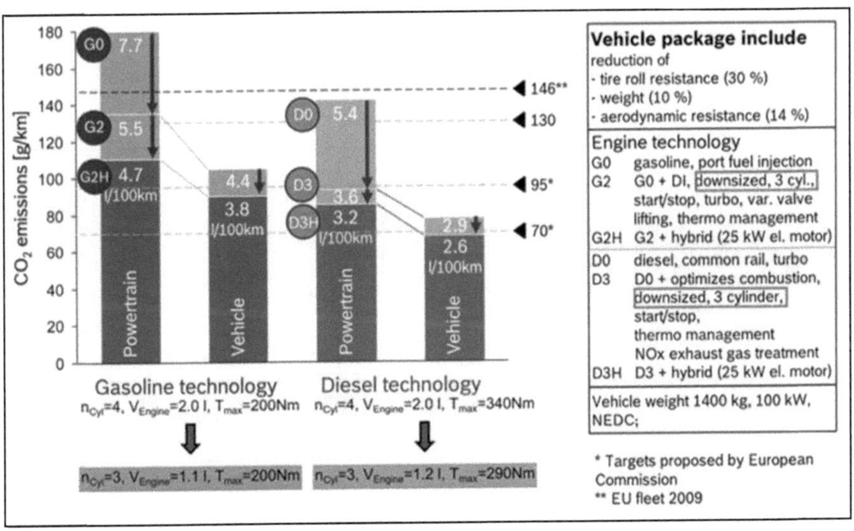

Figure 3: CO_2-Reduction targets of the European Union by Downsizing/ Downspeeding and improved vehicle technology [2, modified]

5.1.2 *Excitation of Torsional Vibrations and Isolation Targets in Automotive Powertrains*

Highly turbocharged combustion engines and especially highly supercharged engines with a low number of cylinders induce unsteady and high dynamic torques in addition to existing high average torques. This might cause reasonable problems with torsional vibrations in the powertrain.

Figure 4 illustrates an increasing engine torque of combustion engines for the period 1990 to 2007. It can be seen that the average engine torque significantly increased within 17 years by a factor of roughly 2.5.

Excitation of torsional vibrations and requirements on torsional isolation of the powertrain are increasing simultaneously in a significant way due to increased supercharging of the engines. Figure 4 illustrates the vibrational excitation and the corresponding isolation target values in automotive powertrains characterized by angular acceleration of the engine's crankshaft for the period 1990 to 2007.

Considerably higher dynamic torques in addition to high average torques and low engine speeds lead to high excitations of torsional vibrations by the engine and can cause acoustical problems as well as strength and lifetime problems of powertrain-components.

In combination with tightening target values for torsional vibrations, requirements on components for reduction of torsional vibrations are increasing at high levels.

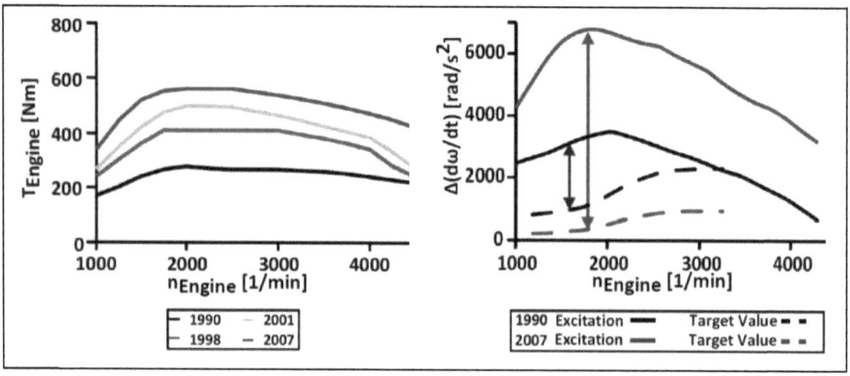

Figure 4: Increasing engine torque caused by Downsizing and Downspeeding (left); excitation of torsional vibrations and trend of isolation target values (right) [4, modified]

Currently automotive technology is moving from conventional combustion engines to hybrid and fully electrical power drives. Low battery capacity of conventional batteries causes small operating ranges and constrains electrical support in hybrid vehicles. The operating range of electric cars is increased by energy recuperation during braking operations and by generating electric energy directly by a generator driven by a range-extender combustion engine. Hybrid and range extender combustion engines increasingly have a lower number of cylinders and are highly supercharged.

5.2 Torsional Vibration Absorber Systems for Automotive Powertrains

Since combustion processes of reciprocating combustion engines cause angular oscillations of the crankshaft, vibration reduction devices are attached in order to improve comfort and to reduce component life adulterating vibrations. Automotive manufacturers usually apply devices like torsional dampers to dissipate vibration amplitudes and isolating units, which isolate the powertrain from the driveshaft. Dampers transform vibration energy into unusable thermal energy. Torsional vibrations isolating units isolate the powertrain from vibrations of the driveshaft.

Recently, centrifugal pendulum-type absorbers draw the attention of engineers, since these attachments improve the performance of conventional vibration reduction systems significantly.

5.2.1 Mechanical Background

Absorbers transform excessive kinetic energy of the vibrating shaft into kinetic energy of partial oscillators which can be transformed back to kinetic energy of the shaft. This results in nearly continuous torque delivery within the powertrain. Figure 5 shows the principle of dynamic vibration absorbers with mass moment of inertia of the drive shaft J and mass moment of inertia of the partial oscillator J_T connected through a torsional spring with spring constant c_T.

The natural frequency of the partial oscillator considered as one-mass-oscillator without damping can be derived:

$$\omega_T = \sqrt{\frac{c_T}{J_T}} \tag{1}$$

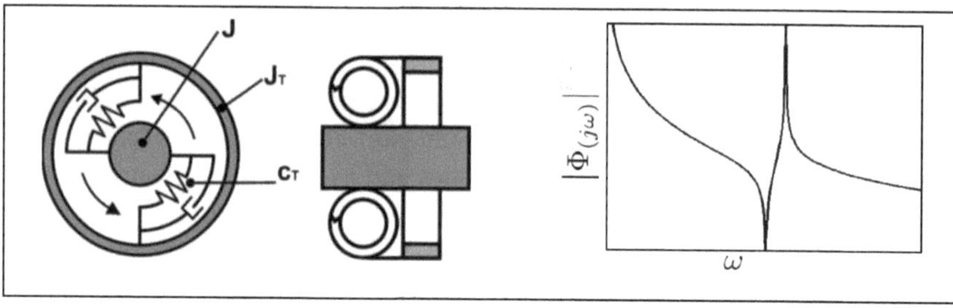

Figure 5: Conventional dynamic vibration absorber and its enlargement function

This natural frequency corresponds to the tuning frequency of the vibration absorber and the partial oscillator absorbs the vibrational energy at this frequency. This can be obtained by the enlargement function (Figure 5).

Centrifugal-type absorbers with several partial oscillators with mass m_T and mass moment of inertia J_T connected to the mass moment of inertia J of the drive shaft with cantilever can be linearized at the operating point of the absorber system (Figure 6).

The linearized centrifugal pendulum-type absorber can be treated as a conventional dynamic vibration absorber with a spring constant proportional to the squared rotating speed of the drive shaft Ω^2 . With a proportionality factor k depending on geometrical parameters the natural frequency of the linearized centrifugal pendulum-type absorber can be derived:

$$\omega_{T,lin} = k \cdot \Omega \quad (2)$$

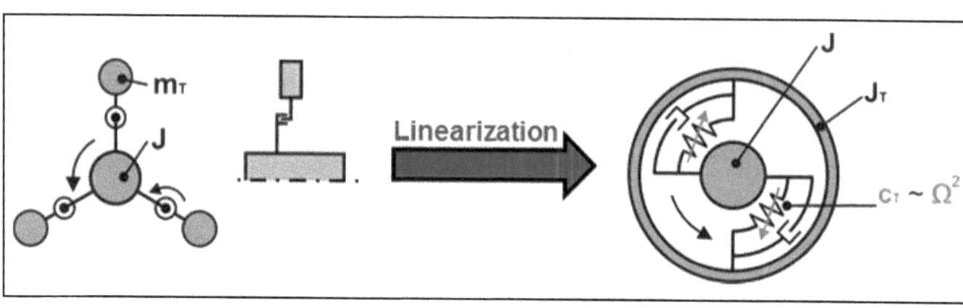

Figure 6: Linearization of a centrifugal pendulum-type absorber with several partial oscillators

5.2.2 *Conventional Centrifugal Pendulum Absorber Systems*

Currently two types of centrifugal pendulum-type absorber systems are applied to automotive powertrains. The SARAZIN-type pendulum absorber is designed as one or several bifilar guided Pendulum Bodies (Figure 7 left). The SALOMON-type pendulum absorber consists of Pendulum Bodies rolling on a roller track on the flywheel (Figure 7 right).

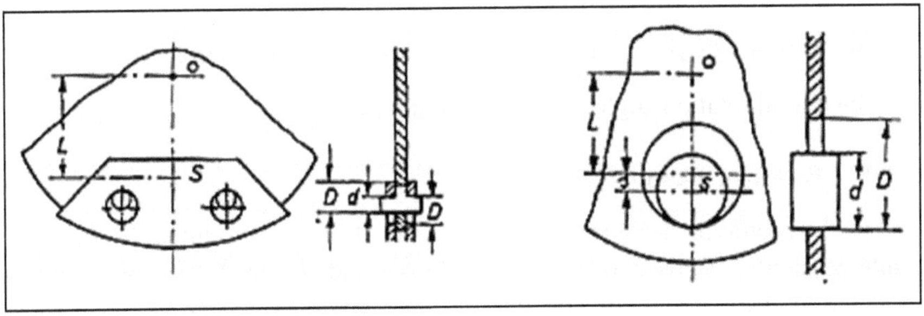

Figure 7: SARAZIN- (left) and SALOMON-type (right) centrifugal pendulum-type absorber [7]

5.3 pCPA – Functional Principle and Prototype-Design

As current centrifugal pendulum-type absorbers are limited to moderate excitations, the pCPA-system enables further Downsizing and Downspeeding of combustion engines.

5.3.1 *Introduction*

There are several approaches for development of the pCPA. Effectiveness of absorber systems is highly limited by friction-caused damping.

At transient operating states, e.g. start-stop-operations, disturbances of the pendulum movement can occur and movement of the Pendulum Bodies can get asynchronous. Additionally, the Pendulum Bodies might strike at deflection limits reaching the maximum rotation angle causing audible noise as well as strength problems of pendulum components.

The ability of absorbing energy depends on the energy storage capacity of the Pendulum Bodies. Hence, the ability to reduce torque fluctuation is restrictively limited by space and target values concerning maximum mass moment of inertia of the absorber unit.

Hence, the pCPA is designed to achieve high effectiveness with minimized friction and optimized performance at transient operating states:

- High effectiveness achieved with minimized friction
- Optimized performance in transient operating states, e.g. start-stop-operations
- Synchronous movement of the Pendulum Bodies
- Prevention of striking of the Pendulum Bodies at deflection limits
- Increased rotation angles of the Pendulum Bodies

5.3.2 Prototype-Design of the pCPA

The pCPA prototype-design (Figure 8) consists of several Pendulum Bodies with tooth gearing, a Carrier with Guiding Links and Toothed Rolling Contours mounted on the Carrier. The Pendulum Bodies are connected by a Unison-Ring. The Smooth Motion-Limiting Device is realized as a torsional spring integrated in the Unison-Ring.

The Rolling Contour carried out in two-piece design consists of two gear segments. The Pendulum Body is also carried out in two-piece design and consists of two segmentally toothed components. The components of Rolling Contour and Pendulum Body are connected by screws and fitted precisely by Fitting

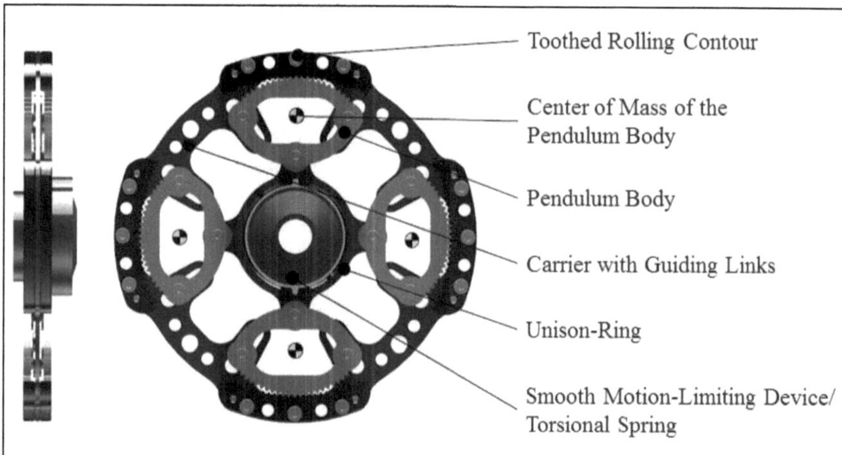

Figure 8: Prototype-Design of the pCPA

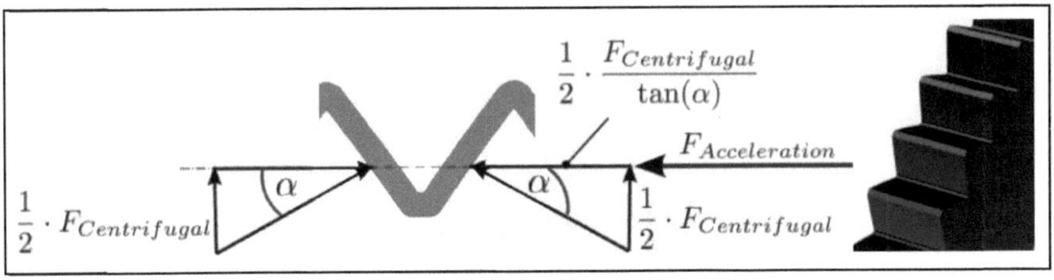

Figure 9: Design of the Gear Toothing

Sleeves. At the contact surfaces to the Carrier and the Unison-Ring Slide Bushes with anti-friction coating are attached. As a result, friction at these contact surfaces is minimized.

Center of gravity, mass moment of inertia and mass of the Pendulum Body is setted by cut grooves, additionally enabling mistuning of the pendulum-type absorber.

The gearing of the pCPA is very untypically loaded. The two functions of the gearing are:

- Guiding of the Pendulum Body
- Supporting the centrifugal force of the Pendulum Body

The gearing is designed as back-lash free gearing with minimized friction. The tangential force on the tooth flank consists of two components (Figure 9):

- Accelerating force of the Pendulum Body
- Tangential Component of the centrifugal force of the Pendulum Body

The Pendulum Bodies are connected and guided by the Unison-Ring (Figure 10) enabling synchronous movement of the Pendulum Bodies. Moment of mass inertia of the Unison-Ring effects the effective moment of mass inertia of the Pendulum Bodies. Hence, the moment of mass inertia of the Unison-Ring has to be optimized. The Unison-Ring is mounted with a plain bearing on the drive-shaft.

The Smooth Motion-Limiting Device is integrated in the Unison-Ring (Figure 10) and mounted to the driving shaft. It is designed as elastomer spring, engaged only when the Pendulum Bodies reach their maximum rotation angle α. With the torsional spring being engaged, the Pendulum Bodies are slightly slowed down and striking of the Pendulum Bodies at deflection limits is prohibited.

Of course, design variations of all components are possible.

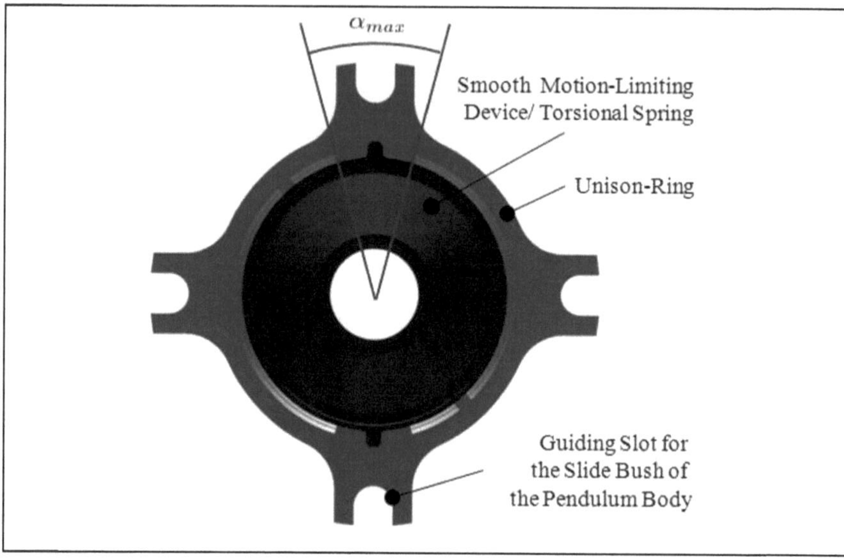

Figure 10: Unison-Ring with integrated Smooth Motion-Limiting Device

5.3.3 Integration of the pCPA into Automotive Powertrains

The pCPA can be easily and space-savingly integrated into automotive powertrains. The pCPA is designed as Add-On Module for conventional Main-Systems (Figure 11).

It can be mounted on the secondary-sided flywheel of conventional Main-Systems or even be totally integrated into conventional Main-Systems. The pCPA can increase effectiveness of Main-Systems reducing the remaining torsional vibrations on the secondary-sided flywheel. A selection of possible Main-Systems is shown in Table 1.

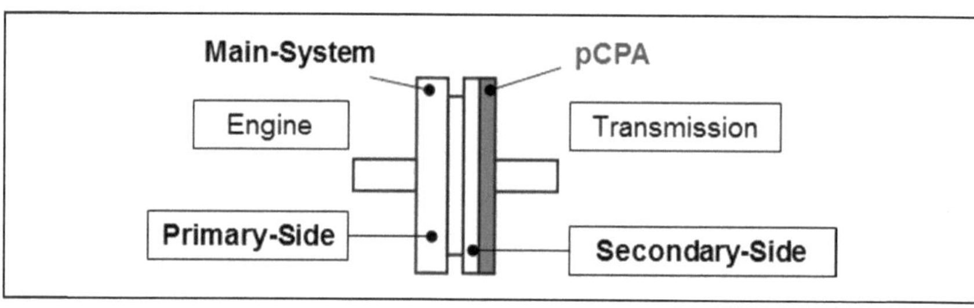

Figure 11: pCPA as Add-On Module for the secondary-side of a Main-System

Table 1: Conventional Main Systems – Selection

Dual-Mass-Flywheel [5]	**Planetary Dual-Mass-Flywheel [6]**	**Hydrodynamic Torque Converter [6]**	**<u>P</u>assive Switching <u>I</u>solator (PSI)**
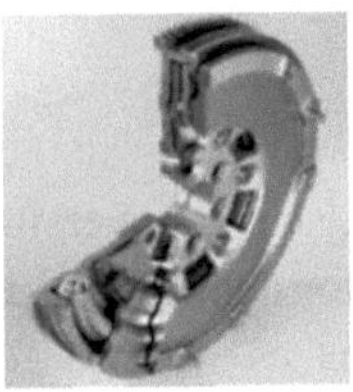		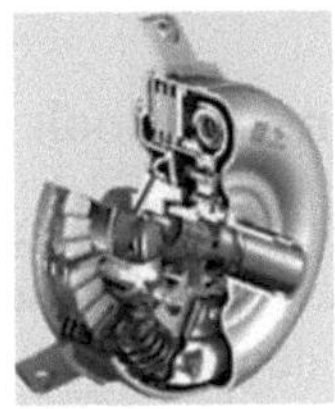	➔VDI conference "Transmissions in vehicles" 2012

5.4 Performance

Based on simulation results considerable enhancement regarding the remaining dynamic torque amplitude on the secondary side of a Dual-Mass Flywheel (DMF) can be determined.

Figure 12 shows simulation results of an exemplary vehicle powertrain with highly supercharged 3-cylinder-engine.

In this simulation the mass moment of inertia of the secondary side of the conventional Dual-Mass Flywheel with integrated pCPA is equivalent to the

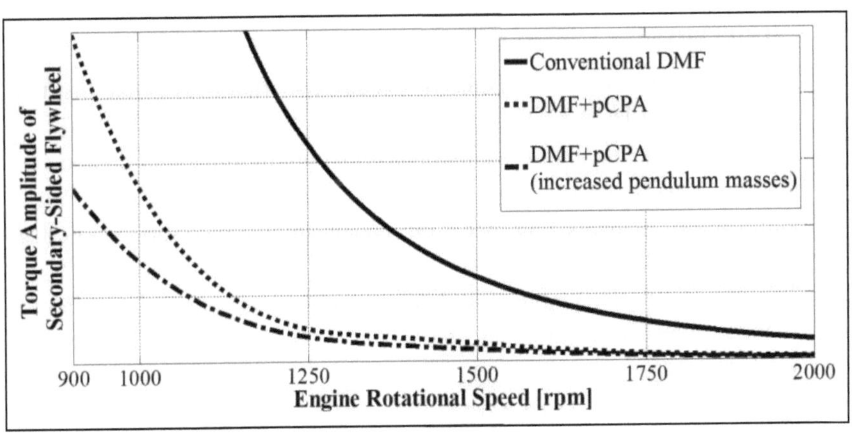

Figure 12: Performance of the pCPA as Add-On-Module to a conventional Dual-Mass Flywheel

mass moment of inertia of the secondary side of the conventional Dual-Mass Flywheel. By increasing pendulum masses the performance of the pCPA can be increased and, in consequence, the dynamic torque amplitude can be decreased further. Performance improvements can be detected especially in regions with low rotational speeds.

5.5 Conclusion and Outlook

The pCPA-concept developed and engineered at the Technische Universität München (TUM) is a new type of pendulum absorber suitable for applications in highly downsized and downspeeded conventional powertrains as well as for hybrid and range extender combustion engines.

The pCPA is designed as Add-On Module for applications with conventional isolating systems and space-saving integration into conventional systems is provided.

The pCPA enlarges the effective range of pendulum-type absorbers and improves the effectiveness and acoustical performance particularly in transient load states and in operational areas with low rotational speeds.

Reduced friction by minimized rolling friction in tooth gearings and small HERTZian pressure on tooth flanks improves the performance of the absorber system. Additionally, performance at transient operating states is improved by a Unison-Ring and a Smooth Motion-Limiting Device.

5.6 References

[1] Hlava, M.; Klobucsar, C. "TomorrowToday", Journal of Austrian Institute of Technology, Issue 1, 2009.

[2] Lenhard, R. "Reducing CO2 emissions with optimized internal-combustion engines", 2011;

[3] Golloch, R. „Downsizing bei Verbrennungsmotoren“, Springer-Verlag, Berlin, Heidelberg, 2005;

[4] Lindemann, P. "Drehmomentwandler – Aufbruch zu neuen Herausforderungen", Schäffler Kolloquium 2010.

[5] http://www.schaeffler.com/ (13.02.2012).

[6] http://www.zf.com/ (13.02.2012).

[7] Hafner, K. E., Maass, H. „Torsionsschwingungen in der Verbrennungskraftmaschine“, Springer-Verlag, 1985.

6 Electro Mechanically Actuated Torque Sensing Multi-Disc Clutch for Automatic Transmissions – Electro Mechanic Powershift Clutch

Prof. Dr.-Ing. Karsten Stahl, Prof. i. R. Dr.-Ing. Bernd-Robert Höhn, Dr.-Ing. Hans-Philipp Otto, Dipl.-Ing. Richard Bäumler, Dipl.-Ing. Christian Hasl

Abstract: The developed powershift clutch system is electro mechanically actuated as required by using a ball ramp mechanism to transform rotational movement into axial load of the disc set. The integrated torque sensor allows control of the clutchs friction torque to realize start-up procedures or continuous slipping operations. By avoiding hydraulics as actuating media and so double energy conversion, it is predestinated for application in electro mobility applications. The developed system is the result of a joint project between the BMW AG, Kirstein GmbH Technische Systeme and the FZG.

Keywords: torque sensor, electro mechanics, multi disc friction clutch.

6.1 Introduction

Automatic powershift transmissions are usually associated with high comfort in longitudinal dynamics and smooth start-up procedures. Along with the continuous advancement of drive train components concerning its effectiveness these transmissions were also in focus. Most evident improvement is a bigger spreading which is reflected by an increased number of gears. In this context the trend of development is to avoid a rising number of clutches per number of gears by using complex transmission concepts as each clutch causes losses which again have to be compensated by other measures. The 8-speed gearbox by ZF (8HP) uses for example as well as their former 6-speed gearbox (6HP) five shifting elements. As a result of downspeeding and other efficiency measures the fuel consumption is no longer necessarily higher in comparison to manual gearboxes. Apart from various improvements throughout the history of powershift transmissions in planetary design, the actuating media of shifting elements hasn't changed since the first automotive application in the early 1950s. Whereas multi-disc clutches within dual clutch or rear axle transmissions occasionally are actuated electro mechanically, within automatic powershift transmissions hydraulic pressure is still used exclusively. Instead of using an oil pump driven directly by the gearbox input shaft, it is possible to use an electric motor to power it as required to decrease dissipation losses. Nevertheless such an electro-hydraulic system causes losses due to the conversion of electric to hydraulic energy.

6.2 Objectives

6.2.1 Objectives of the developed electro mechanic clutch

Apart from the mentioned energy conversion electro-hydraulic systems suffer from unavoidable losses caused by leakages so that the electrically driven oil pump also has to be partially operated when no gears are being changed. Therefore electro-hydraulic systems are not fully demand-actuated in the common sense. Objective of the introduced multidisc clutch was to develop an electro mechanic powershift element which only consumes (electric) energy if the clutch is being actuated. To avoid the use of hydraulic pressure as actuating media a mechanic mechanism had to be engineered. Operating voltage of the on-board power to supply the electric motor is 12 V. The electro mechanic clutch system allows two essential operation modes:

- *position controlled*: The clutch is closed within a parametric defined time period to a parametric defined axial force level. This mode is used to negotiate rotational speed differences. It's used typically to change gears in powershift transmissions.
- *friction torque controlled*: Friction torque of the clutch is controlled using the measured value of an integrated torque sensor of the clutch system. This mode is used to realize continuous slipping – e.g. start-up procedures of motor vehicles.

6.2.2 Objectives of the torque sensor system

Systems of actual series-application use torque proportional values to calculate the controlled operation of the clutch using e.g. the engine characteristics. As several influencing factors vary within real operation of a car, certain inaccuracies have to be taken into consideration using this method. Integration of a torque measuring sensor enables the electro mechanic clutch system to control friction torque of the clutch and allows demand-actuated mechanical load of the systems components.

6.3 Concept

6.3.1 Torque sensor

Strain gauges are commonly used to measure torque by detecting the elongation of mechanically stressed components for special purposes respectively test setups. High expenditure for assembly and a relatively big amount of installation space hinder its usage in series-applications. As an alternative the company NCTE offers the application of wireless torque sensor systems based on the physical phenomenon of inverse magnetostriction.

Figure 1 shows the test shaft for the test rig demonstrator of the presented electro mechanic multidisc clutch applicated with such a sensor system. The shaft shoulder labeled as ‘primary sensor’ was magnetically coded. Deformation of the shaft caused by torsional stress causes a torque proportional variation of the magnetic field which can be measured by the secondary sensor. Application of the patented magnetization is thereby restricted to ferromagnetic metal parts.

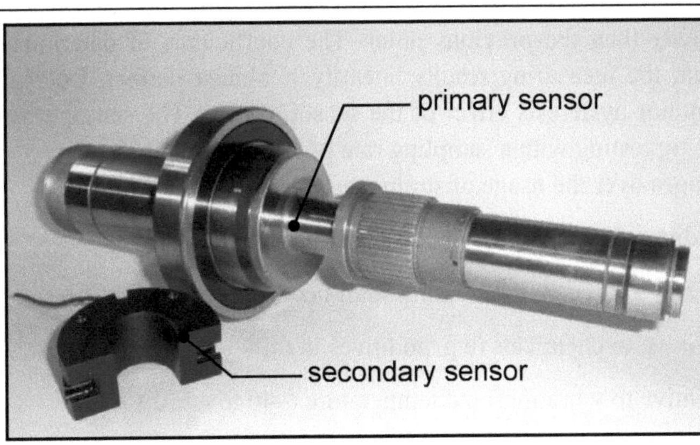

Figure 1: Test shaft with torque sensor

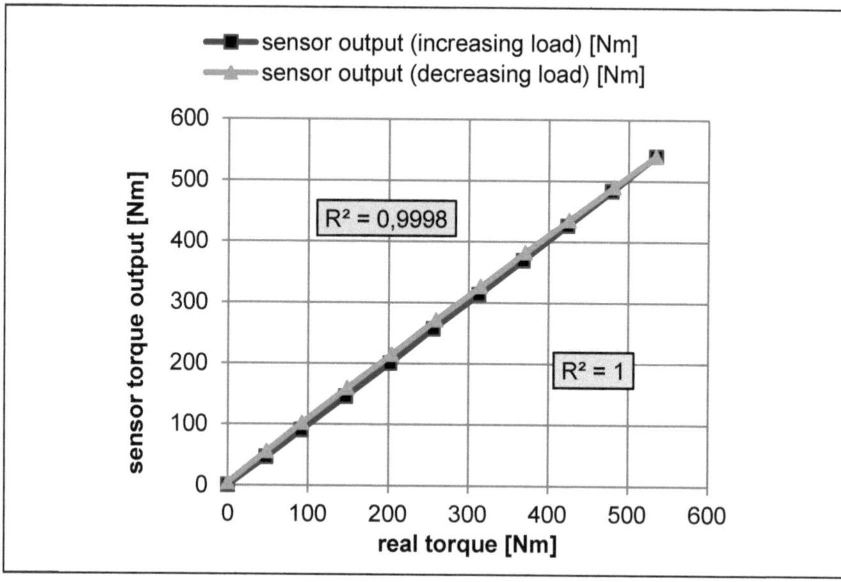

Figure 2: Sensor output quality

Figure 2 illustrates the results of the calibration of the sensor system. The term 'increasing load' means that the load of the regarded measuring point is higher than the previous while 'decreasing load' means that the regarded measuring point is lower than the previous point. The coefficients of determination (R^2) indicate that the measuring results linearity is almost perfect. Decreasing load reveals a minor hysteresis effect of the sensor system. The sensor was operated during test rig testing with a sampling rate of 100 Hz.

Advantages over the usage of strain gauges are:

- no additional parts / mass on the shaft (no imbalance)
- no energy source on the rotating shaft necessary
- insensitive to chemicals (e.g. additives in oil)
- insensitive to vibrations and temperature (–40 to +250°C)

6.3.2 Actuator

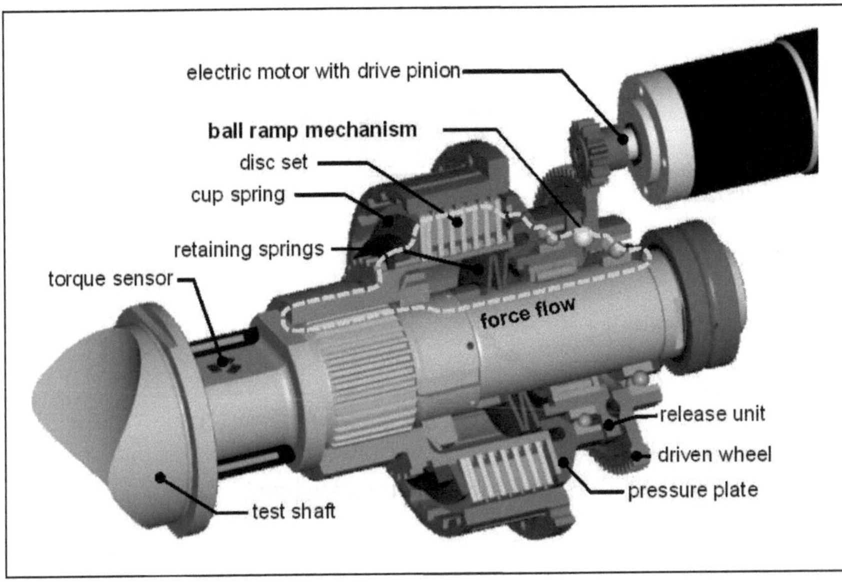

Figure 3: Electro mechanically actuated multi-disc clutch

Figure 3 shows a cut-out of the developed clutch system. To close the clutch the electric motor (12 V) has to be energized. Its drive pinion meshes with the driven wheel and causes the ball ramp mechanism to spread open. Thereby the release unit is moved axially engaging the ball bearing and thus the pressure plate which loads the disc set against the cup spring. As a result axial force in the clutch increases inducing friction torque in the disc set so that rotational speed differences can be synchronized.

The reaction forces of the spreading ball ramp mechanism and on the other side of the disc set are supported by shaft shoulders resulting in a closed force flow circle through the shaft as illustrated in Figure 3.

To gain a linear progression of the axial force over the path, a cup spring was integrated into the force flow. Figure 4 shows the characteristic curve of the clutch system. The data points in the plot symbolized with a rhomb show measured points. Therefor a force measuring ring was installed behind the cup spring. On the y-axis the axial force of the pressure plate (compare Figure 3) is represented over its path on the x-axis. When the ball ramp mechanism is fully

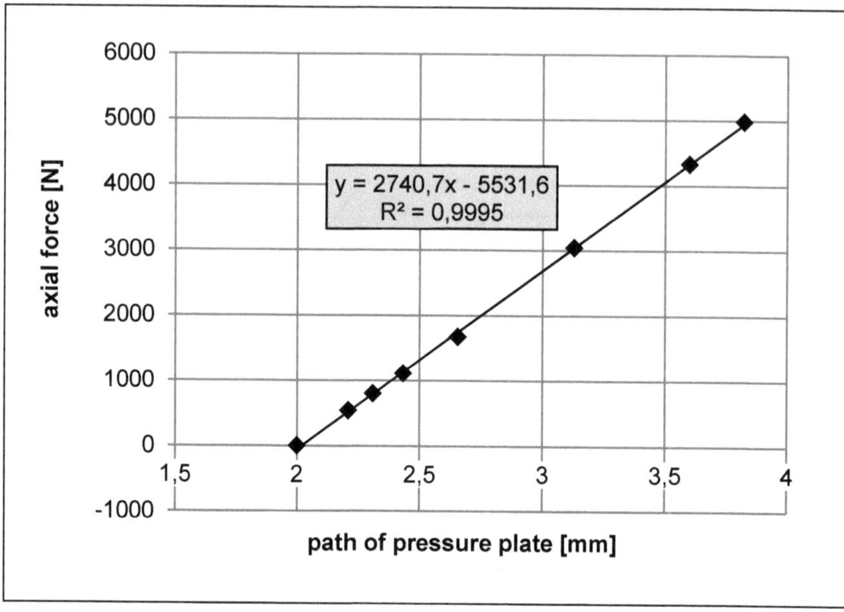

Figure 4: Sensor output quality

disengaged, the pressure plate's path is 0 mm. At a path of 2 mm the pressure plate reaches the 'contact point' where the air gap between the discs is negotiated and further axial movement of the pressure plate causes axial force to rise. Figure 4 indicates a considerably linear relationship between force and path which facilitates a precise controllability of the clutch.

The electric motor is equipped with an integrated brake which enables the clutch system to maintain a designated axial force level without energization of the electric motor. The motor additionally includes a hall sensor which is capable of measuring the position of the actuator with an accuracy of 1/6 rotations of the motor shaft. Considering the systems transmission from the electric motor to the pressure plate axial force can be set with an accurancy of 70 N. The 'overall transmission' is mainly affected by multiple components of the clutch: the gear ratio from the drive pinion to the driven wheel, the pitch of the ball ramp mechanism and the stiffness of the cup spring. Use of an encoder would allow a significant increase in accurancy but was nevertheless in this approach intentionally avoided to prove practicability of the system with the more cost-effective hall sensor.

6.3.3 *Multi-disc clutch*

The concept shown in Figure 3 contains 12 friction surfaces. The design of the prototype which was tested in chapter 6.4 contains only 11 friction surfaces as the axial force flow is led into a force measuring ring which is supported by the housing. For testing, BorgWarner multi segmented discs with paper linings were used. Transmittable torque by the clutch (M_R) depends on the friction coefficient (μ) apart from axial force (F_{ax}), numbers of friction surfaces (z) and middle friction diameter of the discs ($d_{R,m} = 0{,}089\ mm$) according to the following formula:

$$M_R = \frac{1}{2} \cdot d_{R,m} \cdot F_{ax} \cdot z \cdot \mu \qquad (1)$$

Testings with the mentioned discs and a dual clutch transmission oil as lubricant effect a friction coefficient of up to 0,14 which allows a transmittable torque of about 68.5 Nm per 1000 N axial force.

6.3.4 *Control unit*

To keep the development process possibly close to series applications or to obtain a system which can easily be deployed in a trial car, querying of the sensor and activation of the actuator was realized using an embedded system. The project partner KTS designed a control unit based on an Infineon XC2287M microcontroller and programmed the following described software functions to operate the electro mechanic clutch system on the test rig. The units energy consumption in idle is fewer than 5 Watts.

Interaction of the control unit with the PLC (programmable logic controller) of the existing test rig was realized via CAN port. Parameters which had to be varied in the course of the testings were also transferred to the control unit using a CAN-bus PC-Interface.

The contact point of the clutch may be influenced while operation in a transmission system by various factors – e.g. temperature or wear. To compensate these effects it is required to determine the contact point in real time to ensure a defined and comfortable shifting process in any possible operational condition. As it is not intended to equip the electro mechanic clutch system with an axial force sensor for economic reasons and to save installation space, the detection of the contact point has to be realized without use of additional sensors. As an approach therefor the time derivate of the motor current is being used to determine the contact point during each shifting process. When the actuator passes the contact point the slope of this item rises significantly according to an increase of required force to load the disc set.

As the described detection algorithm of the contact point can be executed only during a shifting, to modify following shiftings, the parameter has to be set initially. Therefor a special 'initial detection' function was implemented which for example can be executed once the car is put into operation. To realize this, the system is actuated in a similar way as during the shifting process until the air gap is negotiated and the contact point has been detected. Subsequently instead of loading the disc set, the mechanism is set back to initial position again. This process lasts without detailed optimization about 2 seconds.

The essential task of the control unit is to control the actuator within its two operating modes (compare 6.2.1).

6.3.4.1 Position controlled actuation

For position controlled actuation the software design allows targeting an allocated time for every incremental position measured by the hall sensor. According to the overall transmission of the actuating system from each rotary position results an axial path of the pressure plate and hence an axial force loading the disc set. This correlation allows generation of variable force progressions as a function of time as well as force release within the limitations of the electric motors power.

To demonstrate and test the function of the actuator, a characteristic actuation course was chosen. It represents a typical operational scenario of a powershift clutch and is illustrated in Figure 5. Axial force is plotted over time as well as electric power consumption (dashed line, secondary y-axis) of the control unit which serves the electric motor. The course is divided into 5 sections. Section 1 shows the engaging of the clutch with a set value of 3200 N. In Section 2 the axial force level is hold by the integrated brake of the electric motor as can be seen by the low energy consumption of the system in this range. Section 3 demonstrates an increase of axial force for example due to the demand of a higher power transmission by the clutch. The system being able to vary the axial force of the engaged clutch allows extending the components lifetime by loading them as required. Section 4 shows a partially release of the clutch to the previous force level of Section 2. In Section 5 the clutch system is being disengaged.

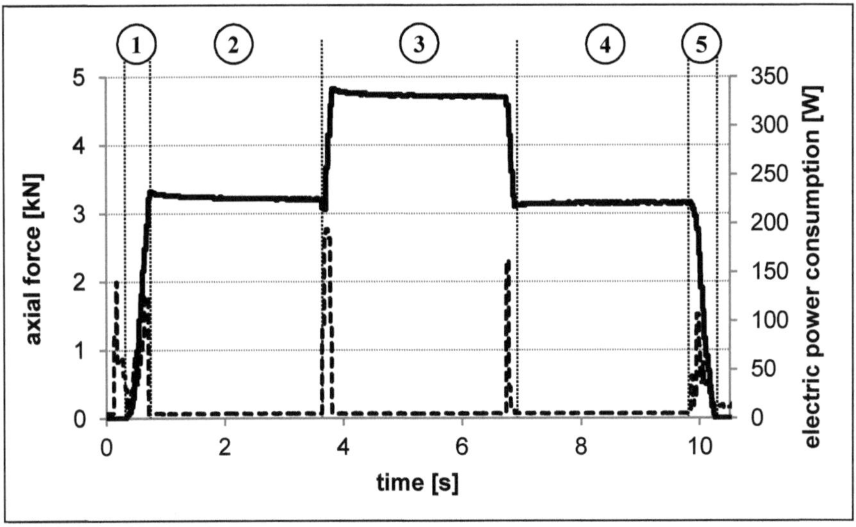

Figure 5: Characteristic actuation course

Table 1 summarizes the variable parameters for the actuation of the shown characteristic actuation course and states the corresponding functions of the electro mechanic clutch system in a potential real transmission system for each section.

Table 1: Variable parameters of the characteristic actuation course

section	variable parameters	representing function in a real transmission system
1	duration axial force level	engage clutch
2	duration	keep clutch engaged
3	axial force level	increase max. clutch/friction torque
4	duration	keep clutch engaged
5	duration	disengage clutch

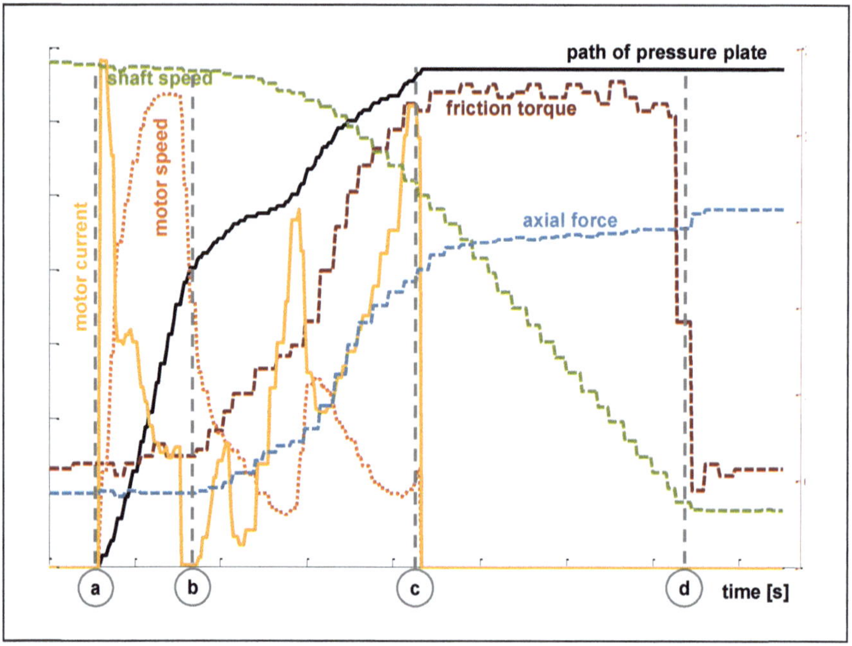

Figure 6: Clutch engagement

Figure 6 shows engagement of the electro mechanic clutch appropriate to section 1 and 2. The shaft is being accelerated by the test rigs drive to a preset shaft speed, before the actuator is engaged at (a). The electric motor is energized at maximum current to negotiate the air gap as quickly as possible. Before reaching the contact point which is labeled as (b), motor speed is reduced to not overshoot it. In the following, according to the preset time requirement, the ball ramp mechanism is spread by further rotation of the electric motor until the set path of pressure plate and for this reason the force level is reached (c). At this point (c) the brake, integrated within the electric motor, is closed – motor current goes to zero. The shaft is being decelerated with constant load and thereby constant friction torque until it comes to stop (d).

6.3.4.2 Torque controlled actuation

The second possible mode of the electro mechanic clutch system allows controlling friction torque in the clutch with the use of the integrated torque sensor system as control variable. For testing, the software allows to set multiple set values of friction torque for defined time periods which are being actuated sequentially when continuous slipping mode of the test rig is activated.

6.4 Testing

6.4.1 Environment of testing

To demonstrate the functionality of the presented torque sensing clutch system a prototype was designed and manufactured to equip a test rig at the FZG with it, as shown in Figure 7. Prior to the testings, a hardware in the loop (HIL) build-up was used for the development of the control unit software at the facilities of KTS (Kirstein Technische Systeme). The aims of the following tests were to prove the functionality and stability of the system under realistic conditions. This includes the use of transmission oil at different temperatures (40–80°C) to cool the multi-disc clutch as well as flywheel masses on the test shaft in the dimension of a cars reduced inertia ($J \cong 1{,}2$).

Figure 7: Test rig at the FZG

The test rig allows two different modes of operation:

- *shift mode*: The testing shaft is being accelerated by the main drive to a desired rotational speed. When this speed has stabilized, a clutch which separates main drive from the testing shaft and the flywheels is opened. By engaging the electro mechanic multidisc clutch the rotational energy is being dissipated until the testing shaft stops.
- *continuous slip mode*: The drive system rotates the testing shaft at constant speed and drive is not disconnected when the clutch is engaged. Due to drive motor power friction torque is applied at contant speed of the clutch.

6.4.2 Static testing

'Static testing' means that the actuation of the clutch is performed without rotation of the testing shaft to avoid any influences caused by dynamic effects with which will be dealt with later.

6.4.2.1 Temperature dependence of the actor

To identify a possible temperature dependence of the actuation process, the clutch system was actuated under identical operational conditions 20 times for each tested oil temperature (40°C, 60°C and 80°C). The measured axial force over time was averaged for 20 actuations of the same temperature. As a result there are no noticeable influences on the axial force application due to variation of the oil temperature identifiable.

6.4.2.2 Variation of rise and decay time

The electro mechanic clutch system allows variable selection of rise time to engage and decay time to disengage the clutch. The following Figure 8 shows test readings of the characteristic actuation courses for four different rise and decay times (300 ms, 500 ms, 1000 ms and 2000 ms). The different plots where triggered to the beginning of section 2 (compare Figure 5). The dashed curves show progress of axial force (primary y-axis) while the continuous lines illustrate the appropriate path of the pressure plate (secondary y-axis). As the contact point is located at a path of about 2 mm, the axial force begins to rise when the contact point is passed. When passing the contact point the slope of the path changes according to the chosen rise respectively decay time. The results illustrate the highly accurate actuation quality of the developed clutch system for variable rise and decay times.

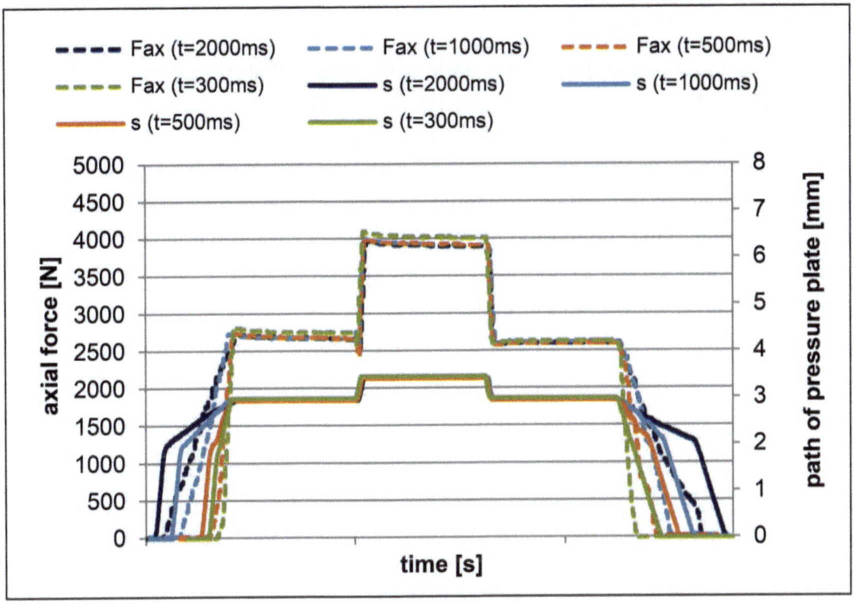

Figure 8: Variation of rise and decay time

6.4.2.3 Variation of axial force level

Apart from the possibility to vary the time until the axial force level is reached, the force level of the clutch can also be varied via parameter within the software. Figure 9 exemplifies that on the basis of measuring results.

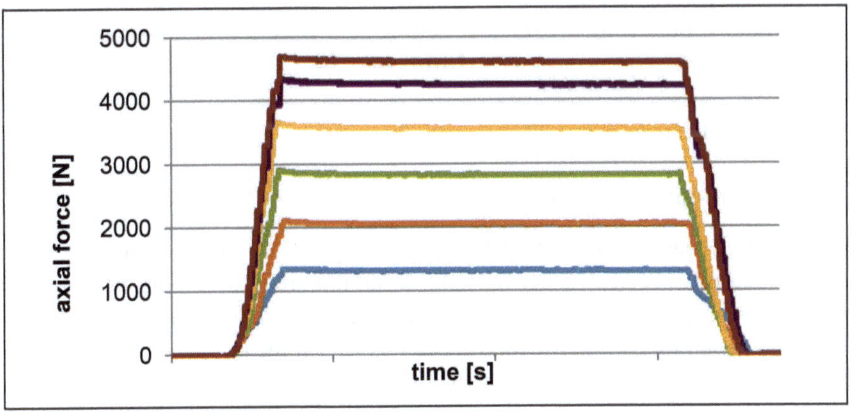

Figure 9: Variation of axial force level

6.4.2.4 Reproducibility of static actuations

To analyze the repeatability of the static actuations, 20 shiftings were plotted on top of each other as shown in Figure 10. Rise and decay time were 2 s each. It can be seen that the actuations demonstrate a high degree of reproducibility. It is possible to repeat the actuation over time with very low deflections. The slightly increased spread at the end of the actuation course is caused by the cycle time of the control unit. This summation error, when timing over serveral seconds, is not relevant for application of the system in a gearbox as the actuations are there initiated instantly by contemporary events.

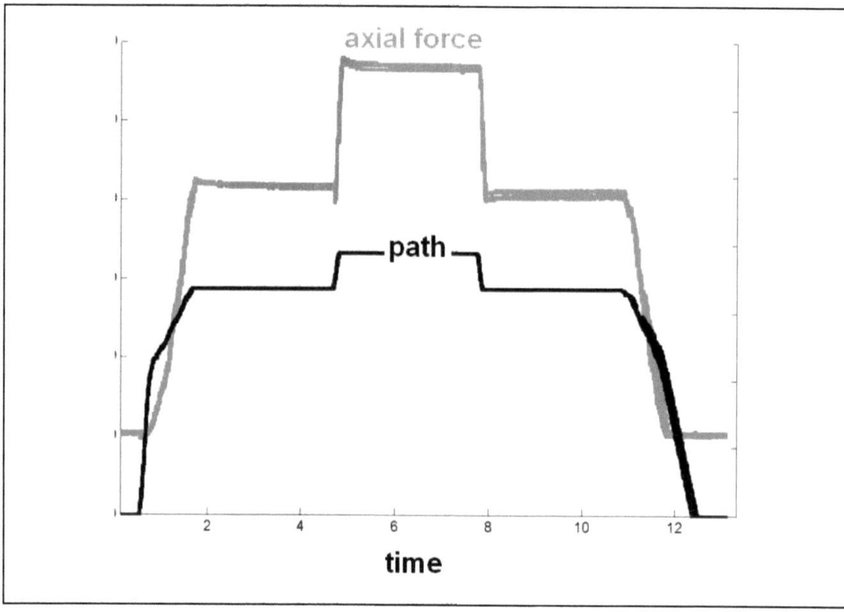

Figure 10: Reproducibility: 20 static actuations

6.4.2.5 Fail-safe operation

It is required that in case of a system error – e.g. failure of the electric motor – the drivetrain is not kept under tension. So an important development objective was to ensure that the electro mechanic clutch opens independently without an active actuation by the electric motor. For this reason the mechanism must not be self-locking. To prove this, the actuator was engaged to several force levels

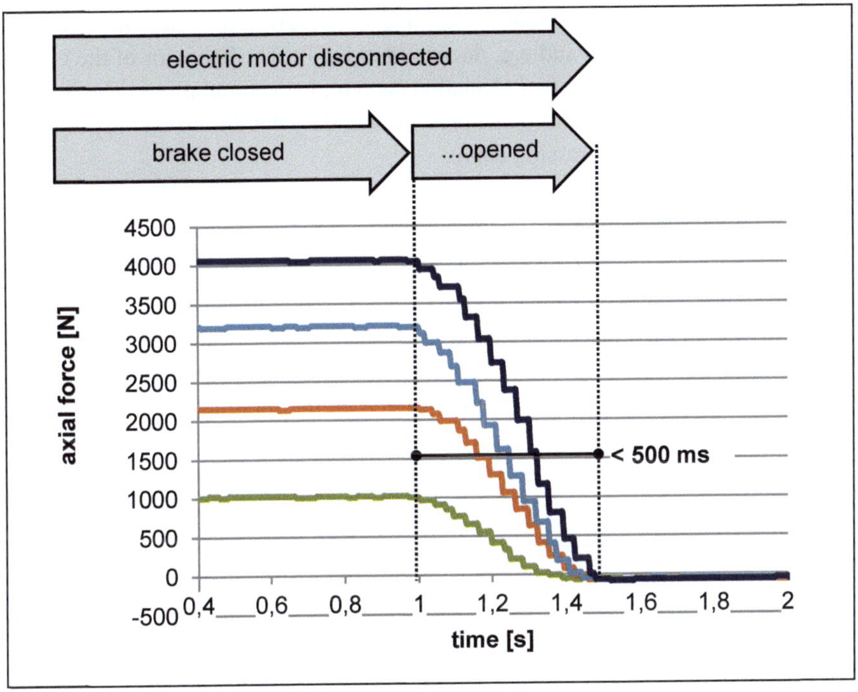

Figure 11: Fail-safe operation

before the electric motor was disconnected. Subsequently the brake of the electric motor was manually opened. Figure 11 shows the result of this test. The spread ball ramp mechanism drives back to initial position when the brake is released and the electric motor is not energized. Relevant for this behaviour are retaining springs illustrated in Figure 3. The results show that regardless the height of the axial force level, the mechanism unloads the clutch within under 500 ms when the brake of the mechanism is released.

6.4.3 Dynamic testing

The dynamic testing, which is described in the following, means that the testing shaft has a rotational speed when actuating the clutch - like in real applications. When engaging the electro mechanic clutch system, the rotational relative speed between inner and outer discs is being synchronized according to the axial load respectively the resulting friction torque. As the axial force sensor is placed behind the disc set, frictional effects in the tooth systems of the discs cause the measured force to be lower than the applied force during the frictional process. This effect is well-known and e.g. documented in [1]. The behavior of the clutch has been verified with analytic calculations and numeric simulations. The time to decelerate is plausible regarding the present boundary conditions and appropriate according to the chosen pressing curve.

6.4.3.1 Reproducibility of dynamic shiftings

Like in static case, repeatability is demonstrated by plotting multiple shiftings into one figure. Figure 12 shows the measured data of 120 sequent shiftings. The test shaft was in each case accelerated to a speed of 1500 1/min and decelerated with a disc set pressing of 1 N/mm². The previously mentioned reduced measured axial force caused by friction effects at the toothings of the discs can also be seen from the figure. The actuator shows reproducibility on a very high level. No single runaways were removed from to plot nor were the shiftings manually chosen. The measured data originates from one uninterrupted testing.

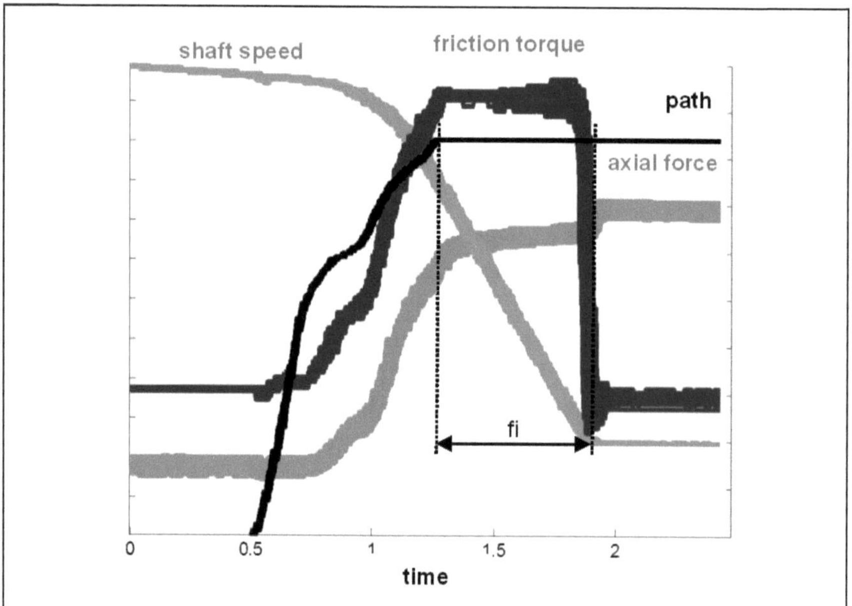

Figure 12: Reproducibility: 120 dynamic shiftings
n = 1500 U/min, p = 1 N/mm²

At the beginning of the labeled section “fi” the constant path of the pressure plate illustrates that the electric motor has stopped and the ball ramp mechanism is hold in this state by the brake. At this point frictional torque of the clutch reaches its maximum. Throughout the labeled section according to a constant pressing the shaft speed is decelerated linear. When the system stands still, the influences by friction on the force measuring ring disappear – the measured axial force complies with the applied force by the spread ball ramp mechanism.

6.4.3.2 Control of friction torque

To test continuous slipping of the clutch with controlled friction torque, the test rig shaft was driven at a constant speed of 84 1/min. Figure 13 shows the measured torque (primary y-axis) and electric power consumption (dashed line, secondary y-axis) over time. The dotted horizontal lines illustrate the two set values which were parametrically set via the userinterface of the control unit. As a residence time for each set value 4 seconds were chosen in this test. The initial torque build-up from (a) to (b) takes place position controlled until 70 % of the set torque value is reached. Subsequently the controller is activated. It actuates the electric motor until the set value is reached with an accurancy of 7 % (b’). In Figure 13 the mechanism is fixed by the brake as the measured torque is within the limit of the 7% tolerance band as can be seen by the low demand for energy of the control unit with an idle power demand of 5 Watts. After the 4s interval

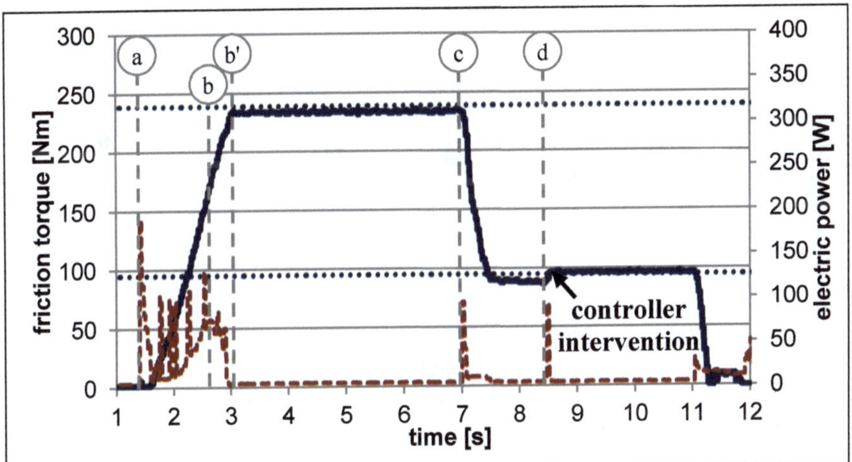

Figure 13: Continuous slipping with controlled friction troque

(b') to (c), torque is being reduced from 240 Nm to 97 Nm. At (d) the measured torque decreases under a value which isn't within the tolerance band. As a result the controller actuates the clutch system to maintain the set value of 97 Nm. The plot of consumed electric power shows a brief peak at this point of time.

As an advantage of this control strategy the energy consumption of the clutch system is kept low. If higher accurancy would be required, the width of the tolerance band may be reduced accepting a higher rate of adjustments by the controller.

6.4.4 Electric energy consumption

An important aspect of the electro mechanic clutch is its demand-actuated operation. Figure 5 shows the electric power consumption of the power supply which serves the control unit of the clutch and the electric motor of the actuator. In the following regardings the electric power consumption was analysed to quantify its potential of efficient actuation. Therefor the known sections according to Figure 5 were differentiated. As the measured value is the electric power for each time increment, it was integrated to electrical work. For each axial force level shown in Figure 14 ten actuations were evaluated. The high-low lines mark the maximum and minimum work performed within a single actuation. The quadrats label the average electrical work for each operating point. The necessary electrical work rises by trend with an increasing force level which is plausible as the counterforce of the cup springs which have to be compressed is the greater the higher the force level is.

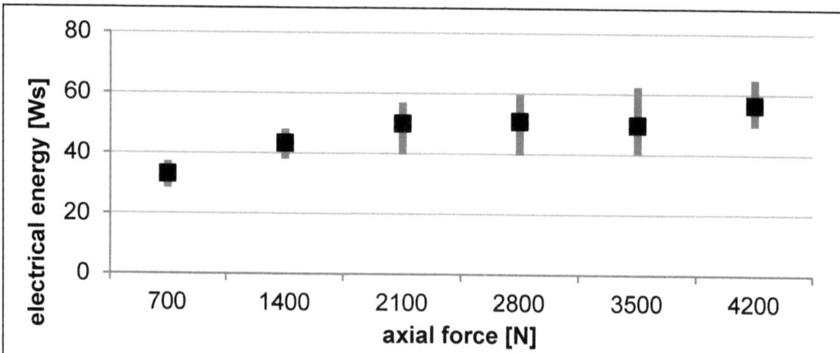

Figure 14: Energy consumption to engage the clutch to different force levels (rise time 500 ms)

Therefor an axial force level of 4200 N allows in a car application with 12 friction surfaces to transmit torque of more than 300 Nm. To engage the electro mechanic clutch to this axial level not more than 75 Ws are necessary. To open the clutch from this force level an electrical work of 55 Ws in average is needed.

Electric power consumption is almost the same for static and dynamic actuations. In futher testings the clutch was engaged to an axial force level of 3000 N within 300 ms rise time once without rotational speed and once at a shaft speed of 1500 U/min. The necessary average electrical work for the static case out of 20 actuations was 32.3 Ws while the result of the dynamic case was 34.8 Ws.

6.4.5 Robustness of the system

During test rig operation of the electro mechanic clutch system all subsystems (actuator, clutch, control unit) proved their stability and robustness. Since the beginnings of the testing 13000 actuations were documented while 2000 were performed including section 3 respectively an increase of axial force (compare Figure 5). Apart from that, numerous actuations during the development process of the control unit, using a hardware in the loop build-up, were performed. During the whole testing no spare parts were necessary and no mechanical errors occurred.

6.5 Conclusion

The demand-actuated design of the developed electro mechanic clutch suits to modern automotive transmission and drive train systems by avoiding the use of lossy hydraulics for actuation. Especially with regard to emobility applications, double conversion of energy (battery → pump drive → hydraulic pump) must be avoided to not reduce the range, caused by limited battery capacity, even more. The presented electro mechanic clutch system allows variable rise and decay times when applying axial load levels which are also variable. Additionally it offers the possibility to adapt the load of the clutch while the clutch is closed – e.g. to increase or decrease the reliability against slipping. As the electro mechanic clutch is solely being actuated by electric components (electric motor with integrated brake) the shifting process easily being adjusted to the respective application via changes within the software.

The results achieved on the test rig prove the systems reliability and reproducibility required for application in a modern transmission system. The spreading ball ramp mechanism allows an even pressure over the entire circumference like a hydraulically actuated clutch which results in a high friction coefficient. Thereby the behavior of the clutch system is as expected similar to its hydraulic counterpart with considerably lower energy consumption. To engage the clutch for transmission of up to more than 300 Nm only 75 Ws are necessary. Common transmission oil pumps consume in best-case scenario (8 bar – 10,5 cSt) at 1500 rpm permanently 400 Watts [2].

The integrated torque sensor allows control of clutch torque by using the realtime measured torque signal as control variable. Therfor a clutch can be used to realize start-up procedures whereby an additional start-up element can be avoided and cost can be saved if the transmission is applicated in combination with a combustion engine.

6.6 Acknowledgements

We kindly thank the funding sponsor Bayerisches Staatsministerium für Wirtschaft, Infrastruktur, Verkehr und Technologie and the project management organization VDI/VDE Innvation + Technik GmbH. Special thanks go to our project partners KTS (Kirstein GmbH Technische Systeme – Augsburg) and to the BMW AG. We also thank NCTE (Non Contact Torque Engineering) for the disposal of the torque sensor system and the associated support.

6.7 References

[1] Winkelmann, S. „Schaltbare Reibkupplungen“. Springer-Verlag, 1985, p. 1.

[2] Conrad, H.; Knauss, R. „Geregelte Pendelschieber-Ölpumpe zur bedarfsgerechten Versorgung von Getrieben“. ATZ. 2011, p. 12.

7 What's the Transmission Content in eMobility

Andreas Kinigadner, Eckhard Kirchner, Rolf Najork

Abstract: Focus of this paper is on the advantages of additional transmission components in electric or hybrid drives for automotive applications. Having described the current challenges for electric mobility, the advantages of a more complex driveline topology are explained in detail by comparing first and second generation vehicles with regard to several customer relevant criteria. Some of the Schaeffler products for eMobility are highlighted with emphasis on their benefits for the global market.

Keywords: Transmission, Torque Vectoring, Electric Range, Efficiency.

7.1 Introduction: Challenge eMobility

The public discussion on future (electric?) mobility has generated additional pressure to engineers developing transportation solutions for tomorrow. Nobody really doubts that hybridization and electrification of passenger and commercial vehicles are required to reduce CO_2 emissions and to decelerate the global temperature rise. But still no market attractive solution for storing electric energy on-board a vehicle exists; battery capacity and price are not yet on an acceptable level.

However, it has become obvious to the public, that generation of "clean" electricity and respective storage is key for success and implementation speed of eMobility. Cost and weight pressure from vehicle manufacturers on the one hand and customer expectations regarding electric range and efficiency on the other hand motivate all major system suppliers to develop their proposals for the market. eMobility has not yet become standard in automotive industry but is already under enormous cost pressure. The diversity of possible concepts will drive the car manufacturers into more alliances for niche models to share development cost between brands. The market challenges to come with the electrification of transportation offers a broad range of opportunities for system suppliers.

7.2 Base assumptions

The prognosis of global energy demand shows a steady increase (see Figure 1), driven out of all sectors, public, industry and transportation. At the same time, indications are clear that the relative amount of fossil energy carriers will continue to rise and by chemistry generate additional emission of CO_2. It is clear, that all emissions for electric power generation need to be taken into consideration when promoting electric mobility.

Another crucial point – one may call it a potential show stopper – is the battery. Currently and for years to come, batteries can be expected to be the most expensive part of the vehicle despite of its limited capacity. Figure 2 shows, that today's typical range of even a B segment vehicle of approximately 400 km would require – based on Li-Ion technology – a 300 kg battery unit for an equal range, not talking cost. The hope of the eMobility community is, that for instance the Li-Air technology proves applicability in industrial and automotive scales in order to reduce the battery mass to approximately 100 kg while allowing practical driving range which are competitive to today's 400 km.

As indicated in Figure 2, the higher degree of efficiency of electric mobility compared to conventional propulsion systems helps to increase the electric range, but energy storage is still the major challenge. Looking to the acceptance

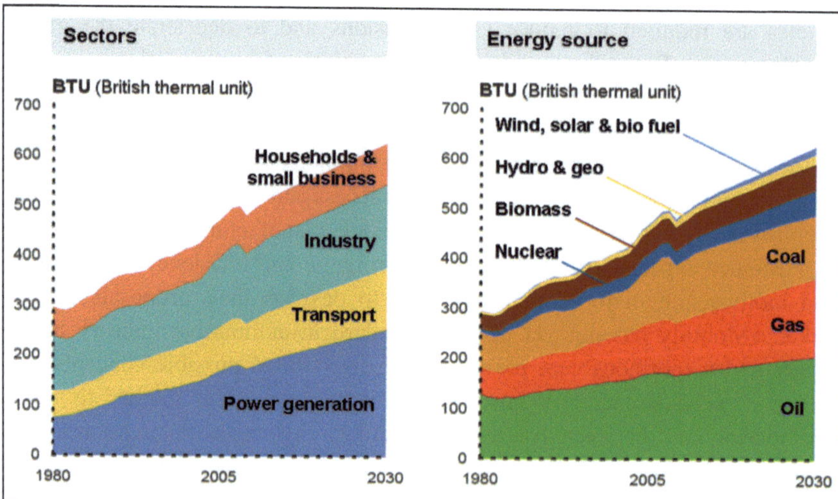

Figure 1: Energy for electric driving – Global energy demand and satisfaction

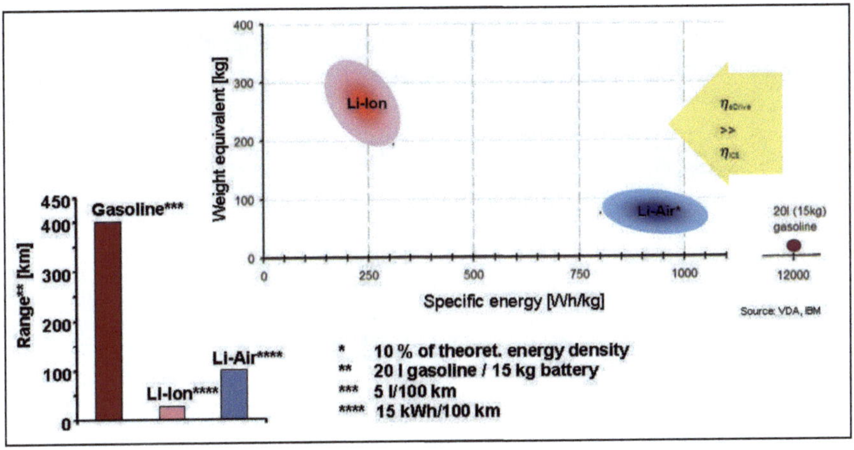

Figure 2: Energy storage in electric vehicles: Comparison of battery versus fuel tank

of eMobility, the price per unit electric power capacity needs to be cut down drastically; a typical cost target across OEMs and system suppliers is in the range of 250 €/kWh. As long as low emission vehicles will have a considerable extra price tag, in most cases buyers will not argue about electrical range. Buyers will choose a conventional vehicle as long as no major political turmoil in the oil producing nations screws the market for crude oil up.

7.3 Transmission influence on electric mobility

To study the benefits of transmission technology for electric mobility a comprehensive study was performed, comparing several first and second generation eMobility concepts in a common framework of a B segment car with same electric motor and combustion engines including respective operational maps etc. to achieve comparability.

As key quantities for the comparison both car manufacturer relevant as well as driver relevant criteria were chosen to compare the powertrain concepts of the vehicles shown in Figure 3. Just by the sheer list of vehicles the necessity of a common evaluation platform becomes obvious. Typical data for the comparison and only to name a few parameters are e.g. vehicle mass of 1250 kg, tire radius 0,3 m, cross sectional area 0,7 m² which are required to bridge the gap between for instance a Mitsubishi i-MIEV and a Tesla Roadster. For the evaluation of

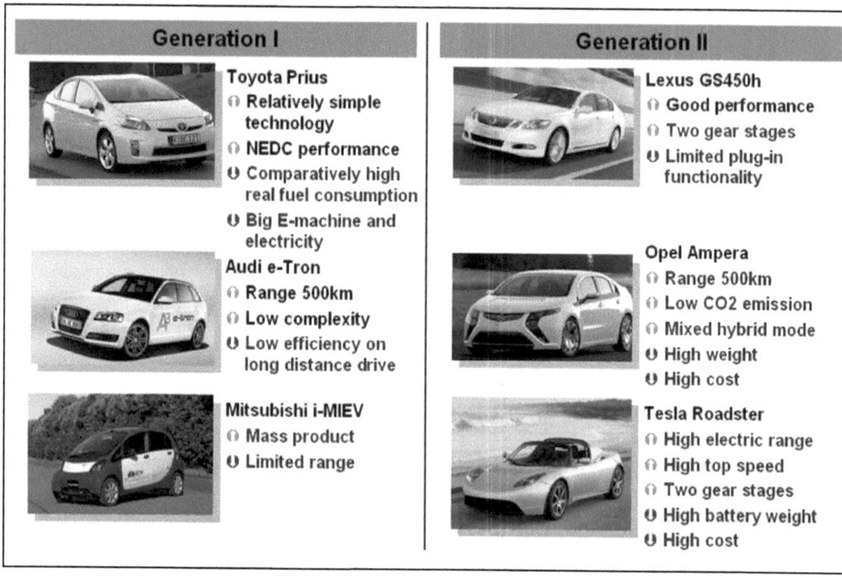

Figure 3: Different vehicles of first and second generation. (Picture source: Internet search)

required combustion engine and electric motor performance key parameters the target acceleration and elasticity were set to uniform moderate targets. Gradeability of all vehicles should be suitable for all-day use, i.e. 25 % at 20 km/h and 5 % at 100 km/h.

Figure 4 displays a comparison of 1st and 2nd generation vehicles with a power split hybrid drive; the drivetrain architecture is chosen similar to the first row of vehicles in Figure 3. It is obvious, that the transmission components integrated into the driveline during system evolution help reduce the requirements on the electric system. The reduction of torque requirements directly influences the motor mass through the smaller required diameter; more important is potentially the reduction of rare earth magnets and the associated cost benefit, which can become accessible to the OEM by adding mechanical components.

The estimated mass increase for the mechanical system balances the benefits of electric system mass reduction; overall, a slight net reduction of the power-train mass can be expected. When looking at the CO_2 emission reduction one can start arguing about an almost neutral business case: a 12 % reduction of the emissions can be roughly calculated to 12 g/km for a good 1st generation concept

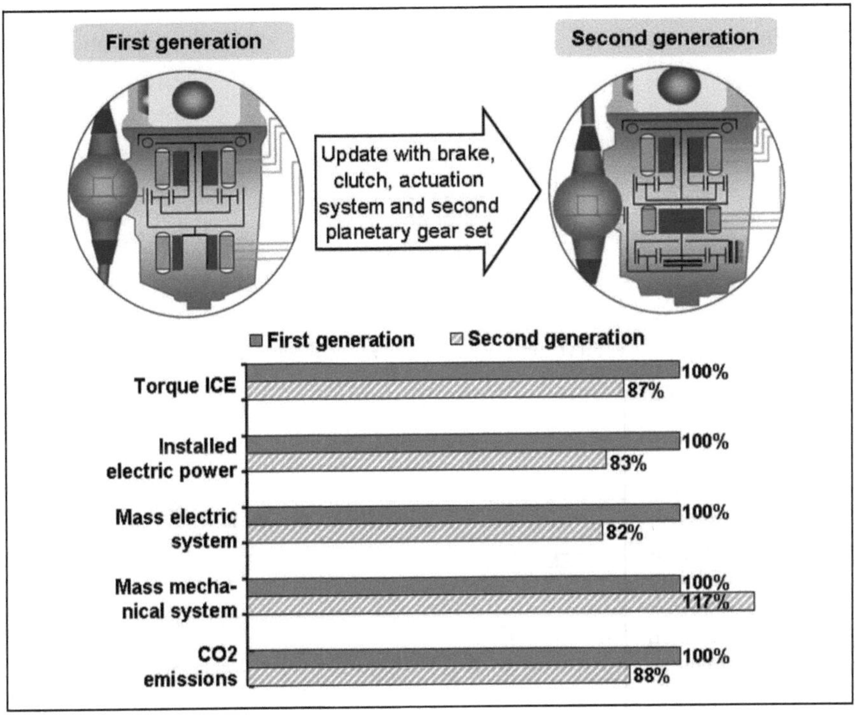

Figure 4: Schematic comparison of system complexity for power split hybrid vehicles and key performance indicators for 1st and 2nd generation concept cars

car which makes roughly a 1000 € span of penalty reduction in 2015. Since power split hybrids are typically designed for a small electric range and no plug-in charging system, the range increase of the vehicle is in analogy to the emission reduction. Daily life usability of the cars increases by the reduction of fuel consumption, depending on the extra price tag for the hybrid system the end customer may calculate promising break even periods for his invest into such a system.

The second comparison of vehicles with regard to transmission benefits is displayed in Figure 5. A simple serial hybrid – which suffers from its inherent deficiency of multiple energy transformation at higher speeds – is compared to a range extended vehicle conceptually similar to an Opel Ampera. Again, the components of the driveline are integrated into the same reference B-segment vehicle as before. Both the combustion and the electric motor can be downsized for the

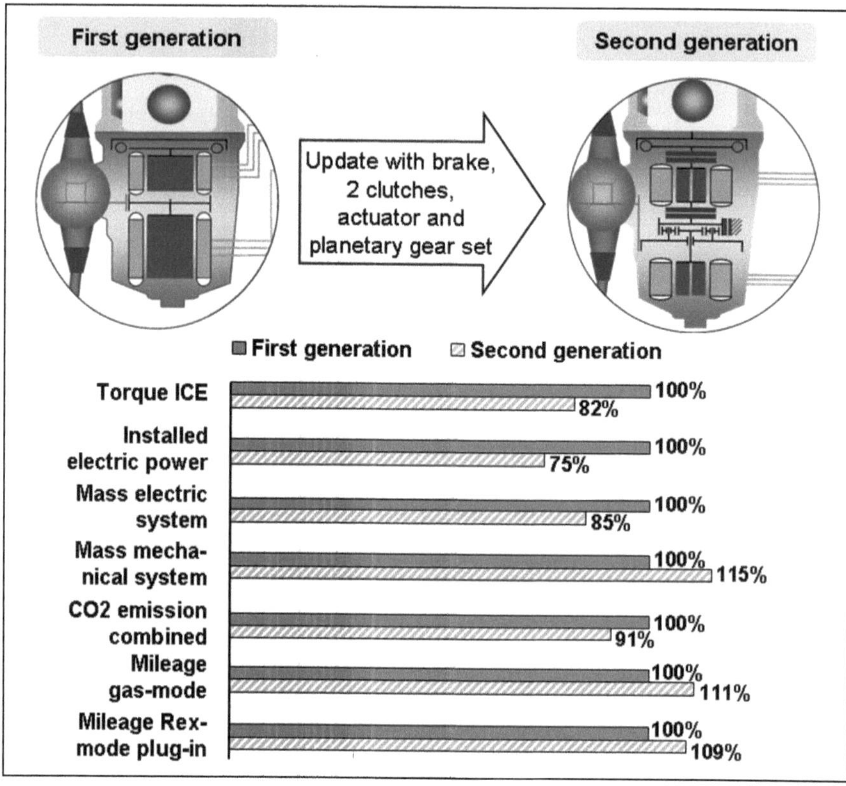

Figure 5: Schematic comparison of serial hybrid and range extender vehicles and key performance indicators for 1st and 2nd generation concept cars

2nd generation concept which is enabled by the integration of a summarizing gear set and a set of two clutches and one break while maintaining vehicle aceleration potential. The different engines can be run more often in efficient operation points. The mechanical link between combustion engine and wheels allows for a considerable fuel consumption reduction at higher speeds. The comparison includes an NEDC for legislative requirements as well as a Schaeffler internal evaluation cycle; it is believed that this comparison is representative for real world driving behavior.

For the purely electric vehicles compared in Figure 6, the most striking improvement is made with regard to the electric torque required to achieve the target acceleration of the B-segment vehicle. While 1st generation is designed

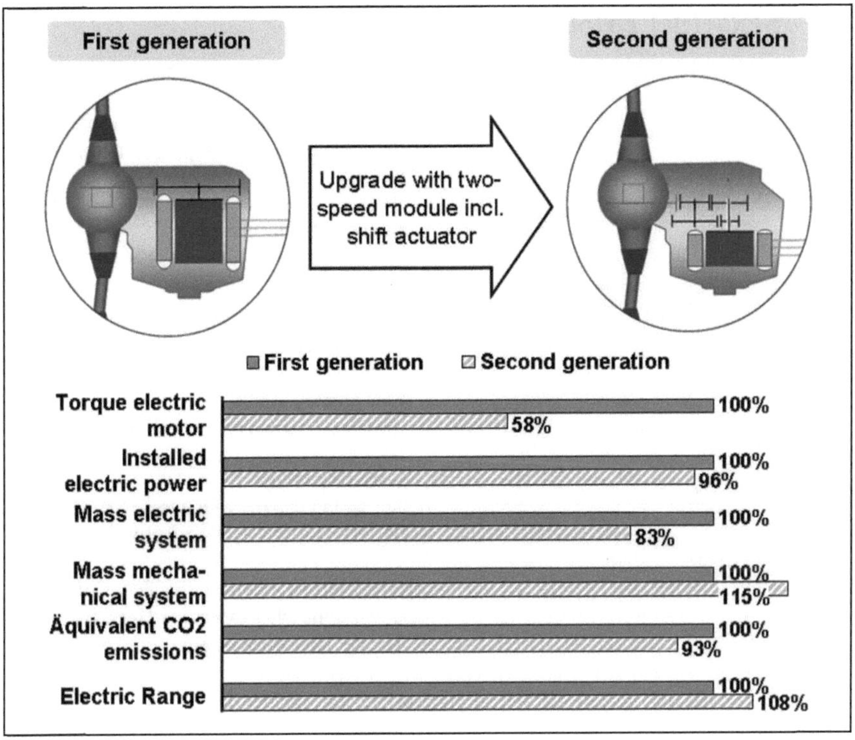

Figure 6: Schematic comparison of pure battery electric vehicles concepts and key performance indicators for 1st and 2nd generation concept cars

with a constant gear ratio, the 2nd generation contains a two speed transmission. The reduction of electric motor torque by 42% for the 2nd generation can be explained by the fact that the torque plateau at low electric motor speed can be used twice, more high torque acceleration is made available. With the reduction of required torque the motor diameter can be reduced and the electric motor can be de-contented with regard to rare earth materials. The electric range, which can be regarded as the most customer relevant criterion, increases by 8 %. Correspondingly, the equivalent CO_2 emissions drop down assuming the German 2010 energy mix for electric power generation; the more fossil material is used for electric power generation, the higher the CO_2 emission reduction effect of the additional transmission components. In the same sense, the relative reduction of emissions will be the smaller the higher the share of regenerative energy production is for Germany.

7.4 Schaeffler eMobility Products

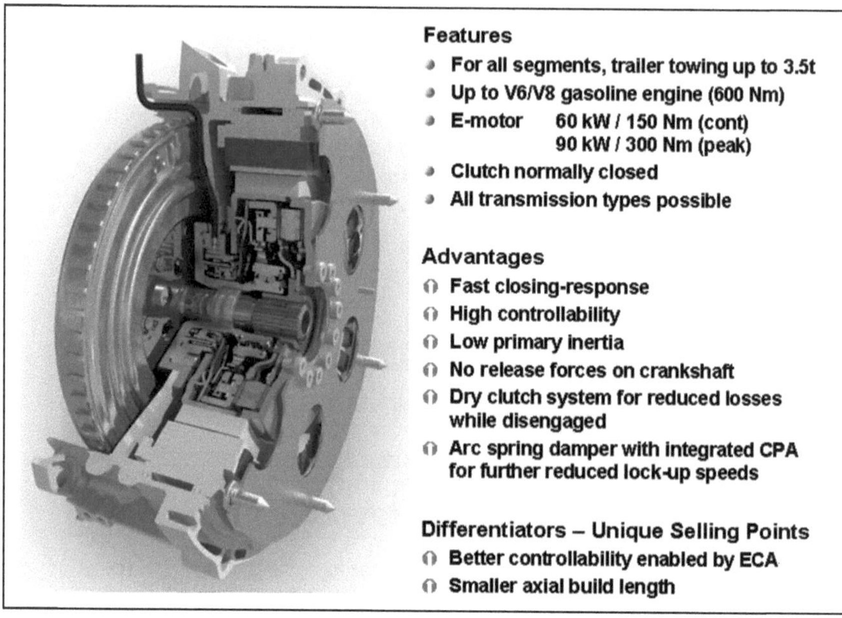

Figure 7: Second generation hybrid module (P2-arrangement)

Schaeffler currently develops three groups of products with respective unique selling points targeting launches within the next years. Figure 7 shows a typical cross section of a second generation hybrid module, the predecessor product is on the market with a German car manufacturer but with a different electric motor. One particular feature of this product is – in addition to its package beneficial overall length – the centrally positioned electric actuator and the integrated centrifugal pendulum absorber. The central clutch actuator achieves benchmark levels of controllability due to an integrated ball screw drive, it is very packaging friendly due to on-board power electronics and controls circuit [1], [2]. The pendulum absorber leads to a very high degree of vibration isolation and almost completely calms down the driveline in combustion engine operation mode. Alternatively, when combined with a hydraulically actuated dual clutch or automatic transmission, the clutch actuator can be actuated out of the hydraulic transmission controls system, requiring one additional solenoid for clutch control only. The integrated clutch can be designed in a comparatively compact way

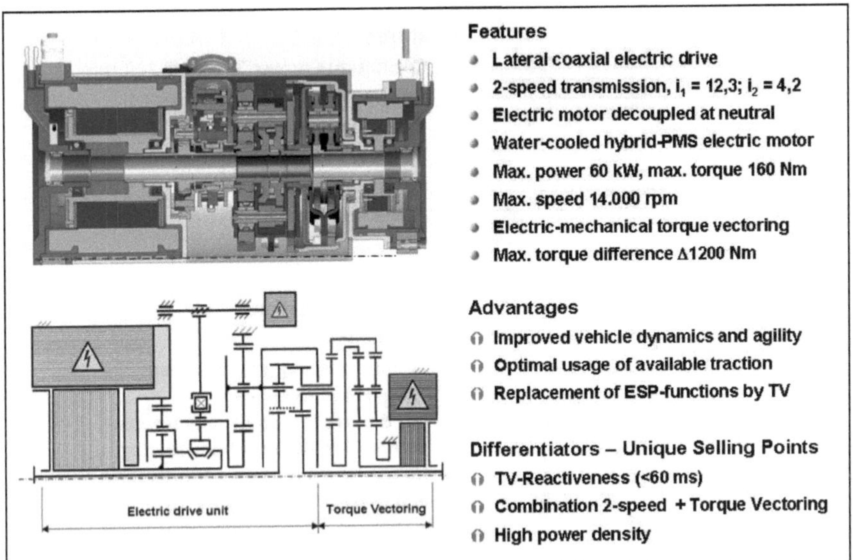

Figure 8: Electric axle module

since it does not need to be dimensioned as a launch device. For higher combustion engine torque levels, a second clutch disc is used in a parallel arrangement, when ICE torque exceeds the 500 Nm limit, a one-way clutch is integrated carrying the drive torque of the ICE while coast torque during recuperation and engine cranking torque are transferred by a single clutch disc. Hence, the modular concept allows for the application to a wide range of engines up to V6 or V8 engines in E-segment vehicles or full-size SUVs with full hybrid capacity across the combustion engine portfolio. The hybrid module is widely applicable to different configurations and supports all hybrid functions such as regenerative breaking or combustion engine start-stop operation.

Another product under development is the electric axle module shown in Figure 8. It features by its modular concept – each function on option – a neutrally open two speed shifting unit, a spur gear based differential and an electromechanically actuated torque vectoring unit. By its design, the rotor of the electric motor is at rest when driving straight ahead, which enables very fast reaction times by simply speeding up the engine by a few 100 rpm and to full torque to achieve full torque re-distribution of up to 1200 Nm. It is even possible to replace a certain degree of ESP functionality by this feature which then really

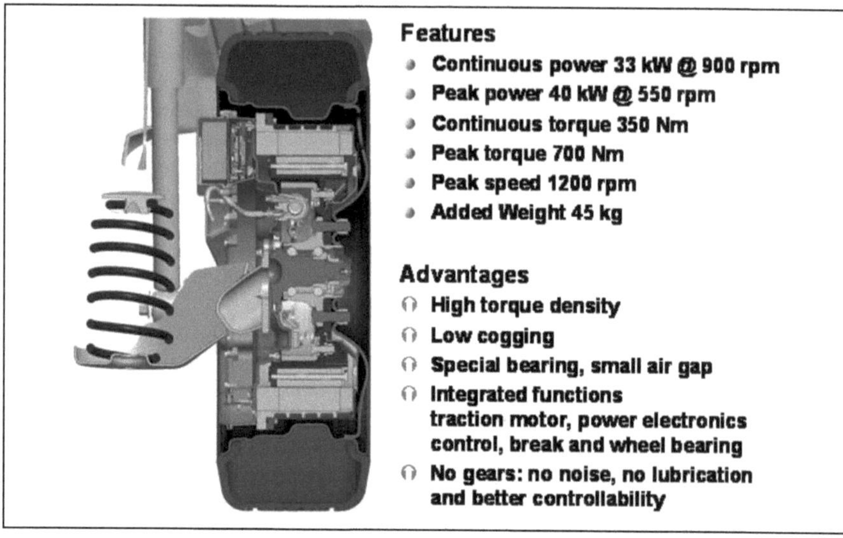

Figure 9: Electric wheel drive

adds energy in critical situations rather than decelerating individual wheels. Since modules are arranged in a coaxial fashion, one can add e.g. the torque vectoring module for more high end sporty cars while offering the potential and benefits of an electric all wheel drive vehicle already to the more volume oriented models [3], [4].

The third group of products under development targets for applications in completely new vehicle concepts, mostly looking at customers in megacities rather than in rural areas. The electric wheel drive shown in Figure 9 will allow for a complete re-design of the vehicle without the standard powertrain compartment, eventually very short front and rear overhang to achieve maximum maneuvering capacity for easy parking in overcrowded cities. It is a technically possible but from a system perspective not ideal approach to combine electric wheel drive at the rear axle with a combustion engine-generator combination at the front axle; one has to cover the cost of a complete electric drive unit in addition to the conventional powertrain without exploiting the possible features of a new vehicle concept.

The battery for the electric wheel drive can be hosted e.g. in a sand-which floor configuration, the wheel hub drive unit will also host the power electronics unit. The very short communication path open possibilities for e.g. co-operative breaking systems which by simulation lead to a reasonable reduction of breaking distance for front-axle located electric wheel drive systems [5]. When installed at the rear end of the vehicle, the improvements in driving stability by incorporation of ESP-functionalities can be named as a major benefit. Moreover, Schaeffler has demonstrated with a demonstrator vehicle that unbound masses do not negatively influence rear axle vertical dynamics when equipped with a electric wheel drive module. However, it is important to keep mass increase due to motor, electronics etc. on the wheel hub under strict control, if the mass target given in Figure 9 is drastically exceeded, vertical dynamics will suffer.

7.5 Conclusion

The benefits of transmission technology in hybrid and electric vehicles are shown in this article based on an analytical 1:1 comparison of powertrain configurations. Based on the data shown, a 10 % increase of electric range can be realized by improved transmission technology in the powertrain with still the same battery capacity. However, since volumes are small and development cost are extremely high, the new energy vehicles will suffer unless governmental programs will push the electric vehicle growth rate with incentives. Another potential game changer can be the oil price: a sudden rise can be driven by a political turmoil in the middle east at any time. Higher fuel cost will reduce the break even period for private customers and could speed up the electrification process. Same holds for the battery technology: A break-trough in energy storage could accelerate the growth rate as well when high-capacity low-cost battery technology becomes available.

With even more concepts than shown under development, Schaeffler is well prepared for upcoming challenges in eMobility. The authors are convinced, that their products ranging from component to propulsion systems offer unique benefits in performance, efficiency and light weight technology for electrically driven transportation. With the modular approaches shown, technically and commercially attractive solutions become realistic ranging over several vehicles segments but utilizing common components. Of course, more attempts with regard to standardization – also looking for partnerships – will be required to cut down cost for the end customer to make electric mobility more attractive. Without economy of scale effects pushing down cost, the desire of the governments for more climate friendly transportation will remain a vision.

7.6 Acknowledgements

The authors take the opportunity of acknowledging several contributions of their colleagues: Martin Dilzer performed most of the calculations with his team used for the comparison and Sebastian Wielgos took the challenge of proofreading the manuscript.

Finally, the tremendous support of the Schaeffler Group for the newly established System Division eMobility needs to be mentioned.

7.7 Bibliography

[1] Schneider, H.-J. "Hybrid Module for Automatic Transmissions with Torque Converter", 10th International CTI Symposium Innovative Transmissions and Hybrid & ELectric Drives, Berlin 2012.

[2] Wagner, U.; Reitz, D. „Die Zukunft kommt automatisch – Effiziente Automatikgetriebe als Basis für Hybridisierbare Antriebsstränge", 9. Schaeffler Kolloquium, Baden-Baden, 2010.

[3] Smetana, T. "Electric Drive with integrated electric-mechanical Torque Vectoring", 10th International CTI Symposium Innovative Transmissions and Hybrid & Electric Drives, Berlin 2011.

[4] Vahlensieck, B. et al. „Erste Erfahrungen mit dem Electric Vehicle Drive von ZF", VDI Symposium Getriebe in Fahrzeugen, Friedrichshafen, 2012.

[5] Poll, D. „Elektromobilität: Renaissance des Radnabenantriebs", Produktion, (45) 2011.

IV Vehicle Concepts

8 Impact of Electrification on the Vehicle Concept – Potential of Determining Components and Technologies

Dipl.-Ing. Johannes Fuchs, Dipl.-Ing. Thomas Riemenschneider, Dipl.-Ing. Rainer Huber, Prof. Dr.-Ing. Markus Lienkamp

Abstract: The design of electric car concepts requires the development and integration of new components that have not been relevant in the design of conventional combustion-driven cars. High-voltage battery systems, electric powertrains, fuel cell systems and hydrogen storage systems for instance, are challenging with regard to an efficient package and design of constructed space. In addition, compromises between the functional and geometrical layout of these components, which determine the overall vehicle concept, are necessary. In respect of development of future electric vehicle concepts it is essential to know the interdependencies between the entire vehicle, its determining components and not least their underlying technologies. The first part of this paper presents the key components that determine the concepts for battery electric vehicles (BEV) and fuel cell electric vehicles (FCEV). Using the example of the BEV traction battery system, an approach for design and analysis of electric key components for a basic vehicle concept is introduced afterwards. In this context, a process model as well as dimensioning and construction tools based on MATLAB/ Simulink and CATIA V5 have been developed and are to be explained in this paper.

Keywords: vehicle concept, battery electric vehicle, fuel cell electric vehicle, vehicle architecture, package.

8.1 Introduction

8.1.1 Initial situation and problem outline

For the development of electric and hybrid vehicles, design and integration of new specific components are of particular importance. High-voltage battery systems, electric drivetrains, fuel cell and hydrogen storage systems for example require efficient packaging as well as a suitable functional design to develop coherent complete vehicle concepts. Similarly, compromises have to be found between the functional and the geometrical layout of the key components, which have a decisive influence on the vehicle architecture and the complete concept.

Moreover, there are increasing dynamics in the further development of the components mentioned and their underlying technologies, of which the influence on prospective complete vehicle concepts has not yet been sufficiently determined. Components and technologies that are taken as the basis for electric vehicle concepts today require substantial development effort to meet the demands of prospective vehicles [1].

This poses new challenges in the development of electric vehicle concepts. However, in this respect it is necessary to coordinate the requirements of all vehicle disciplines in order to create a cohesive and in many aspects optimised concept draft.

8.1.2 Objective

Knowledge of the interactions on the technology, component and vehicle concept levels plays a central role for prospective vehicle concepts. The interactions of these three levels (see Figure 1) in the field of electric vehicle architecture will be examined. Furthermore, implications for concept development of electric vehicles will be derived. On the one hand, information on feasible vehicle concepts on the basis of underlying technologies will be worked out. On the other hand, requirements for applicable components and technologies for intended vehicle concepts can be derived.

The consideration of this objective against a chronological background is a fundamental challenge in this regard. As described in Section 8.1.1, important

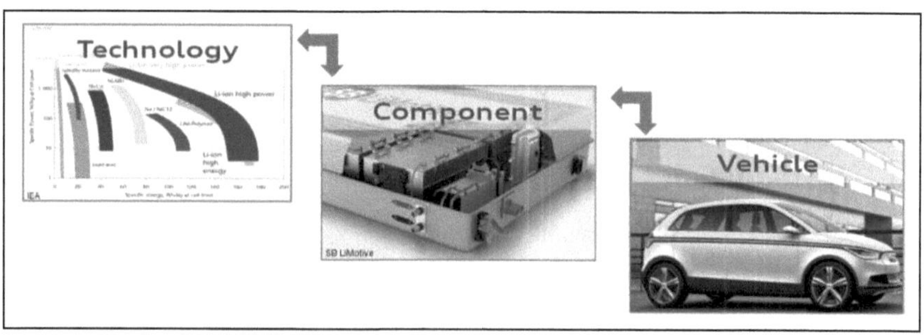

Figure 1: Relevant levels of analysis: Technology, component and complete vehicle

levers for the configuration of future vehicle concepts can be found in the development of key drivetrain and energy storage technologies. For this reason, appropriate technology roadmaps and scenarios are to be compiled. These support a time-discrete analysis of the development in the fields of technologies, components and complete vehicle concepts (see Section 8.4.2).

8.1.3 Terminology

The complete vehicle concept in conjunction with the vehicle package involves (according to [2]) styling, aerodynamics, safety, drivetrain and chassis concept, dimensional layout, production, service, legal requirements and the costs of the complete vehicle. Thus, the engineering merging and evaluation of different requirements and solutions for verifying the feasibility of vehicle drafts emerge as the essential task in the concept development department. Alongside this, the term "vehicle architecture" is also used in the context of vehicle development, meaning a description that is not universally valid or even separated from the "complete vehicle concept". Vehicle architecture as defined by the project presented in this paper comprises the packaging and integration of key components, the rough dimensional layout as well as the physical vehicle structure and the morphology of installation space. The features of platform or modular design are also included. The topology of the drivetrain and of the energy storage represents a further fundamental part of the vehicle architecture (see Figure 2).

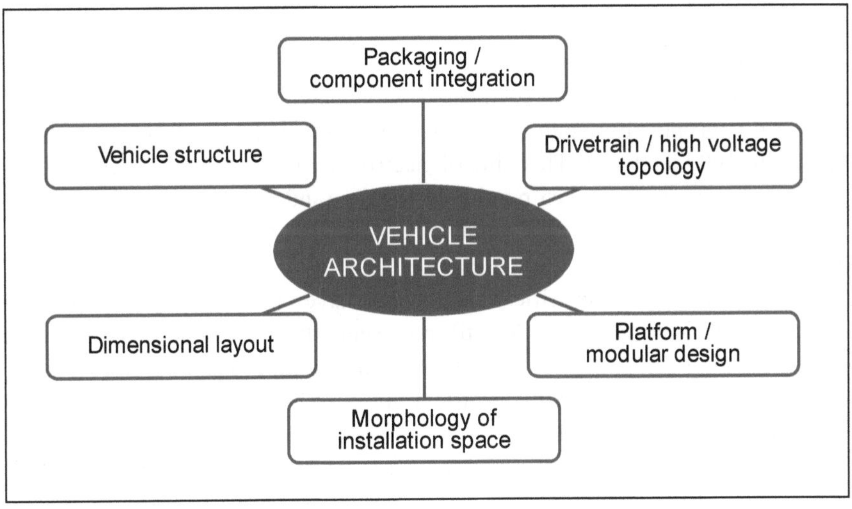

Figure 2: Domains of vehicle architecture

In this regard, electric vehicle concepts in particular offer a huge variety of possible configurations. These have an immediate impact on the architecture and the complete concept of a vehicle.

8.2 Installation space and packaging of the key electric components

8.2.1 Concept determining components of electric vehicles

The number of new components and their specifications are dependent on the level of electrification, which can be defined as the contribution of the electric drive system to the total propulsion. This is shown in Figure 3 where the different levels of electrification are qualitatively classified with reference to a chronological scenario. Without going into details concerning specification and layout of different hybrid and electric drive concepts, reference is made to the appropriate literature, e.g. [1] or [3].

For implementation of micro hybrid concepts, the drivetrain itself is not affected as the electrification measures are restricted to the auxiliary units. According to [4], this concept does not describe a hybrid drive as there is no contribution to the vehicle propulsion. Electrical involvement in the drivetrain first comes with the mild hybrid layout. Redesign of the starter-generator enables electric boost and recuperation function. Full hybrid vehicles provide appreciable electric driving, even if for limited ranges of only a few kilometres. Plug-in hybrid concepts are characterised by the facility for the external recharging of the traction battery. In case of an adequate energy content of the battery, these plug-in hybrid concepts allow purely electric driving for daily use distances of more than 50 to 100 kilometres. This kind of electrification is extremely interesting for vehicles that are to be driven purely electrically (and thus with zero local emissions) over short distances and which are also used as a fully-fledged vehicle for long-haul mobility [1].

All aforementioned electrification concepts have one aspect in common: The internal combustion engine (ICE) still represents the main concept-determining and dominating component in the vehicle architecture. Frequently discussed degrees of freedom in design of electric vehicle architectures are only effective when the ICE and all its associated components – such as the fuel supply, exhaust system and mechanical drivetrain – are no longer in the picture. For the

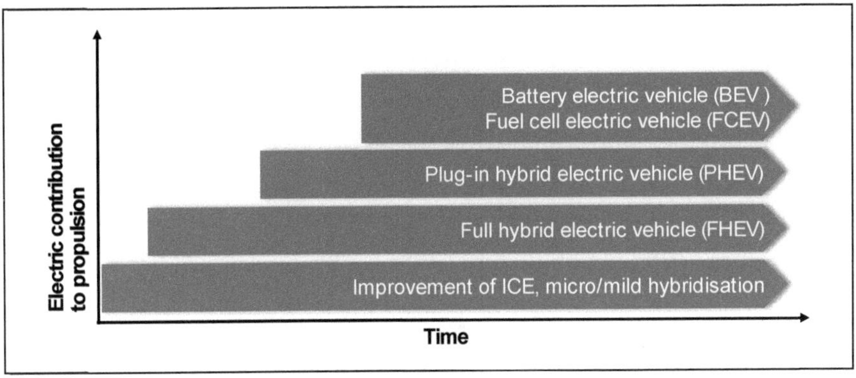

Figure 3: Levels of electrification

reasons described above, this paper purposefully focuses on pure electric drive forms. This places the focus on the research of battery electric vehicles (BEV) and fuel cell electric vehicles (FCEV).

Using suitable methods such as influence analysis [5], the most important interdependencies between components among each other and the dependencies between the components and the complete vehicle characteristics can be demonstrated. Figure 4 shows, as an example, the result of such an analysis of the interactions between the components of a battery electric vehicle in an influence port-

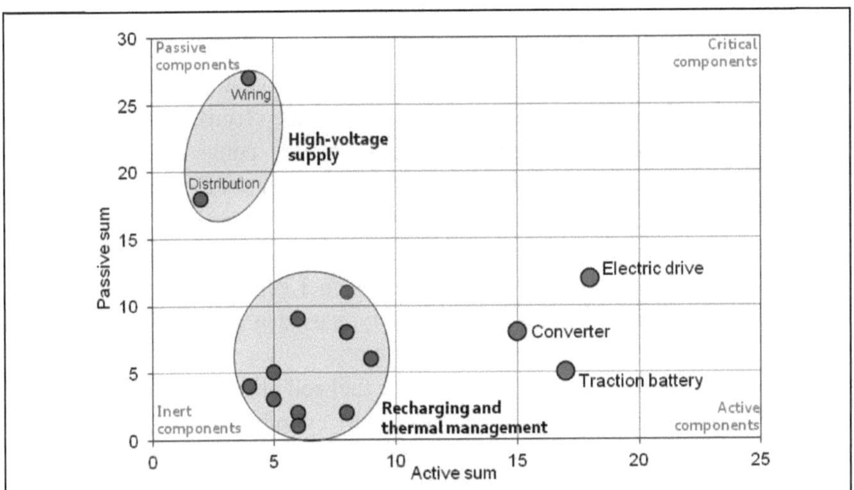

Figure 4: Influence portfolio for concept-determining BEV components

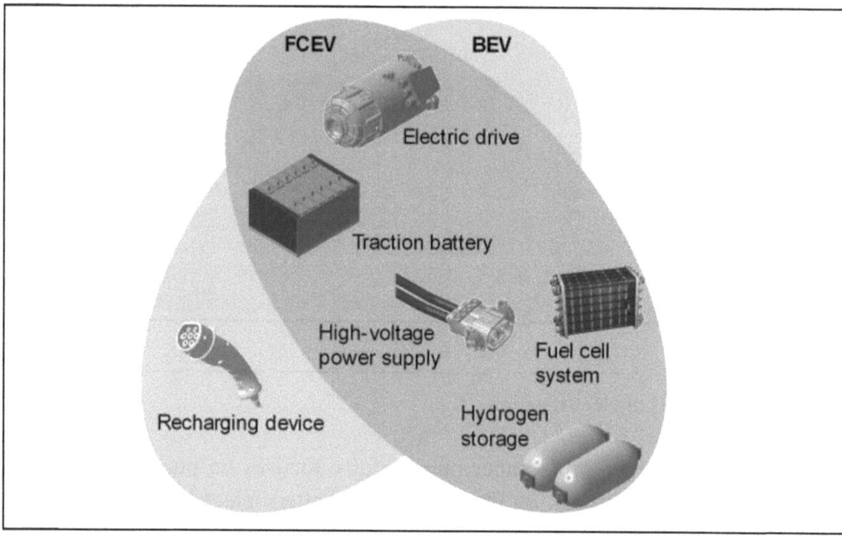

Figure 5: Concept-defining electric components

folio. It can be seen that basically drivetrain and energy storage components have a strong influence on the rest of the high-voltage system. The high-voltage supply is heavily influenced by this, so that it is to be considered as a cost- and weight-driving factor in the early concept stage. Variants of battery charging methods provided are also reflected in the high-voltage architecture.

The systems illustrated in Figure 5 with their corresponding individual components represent the main concept-determining components of purely electric-driven vehicles. The electric drive system includes the electric machine as well as a single-speed or multiple-speed gearbox and the converter. The traction battery system, which is assembled from single cells and modules, is used for energy storage. The high-voltage supply provides the networking of all electric components. A recharging concept comprising a charging unit and interface is not required in FCEV concepts, which use the fuel cell as the primary energy supply for propulsion. Here, the traction battery is used as a power cache for load peaks as well as for storing recuperated energy.

Hydrogen is the primary energy carrier for fuel cell electric vehicles. It can be stored in the vehicle in different ways. Storage by cylindrical high-pressure tanks (at up to 700 bar) is state-of-the-art and represents the most favoured solution of carmakers. There are also solutions for liquid or hydrid storage. Geometrically freeform tanks are still in the research stage.

8.2.2 Assumptions for topology of installation space

In the field of alternative drive and vehicle concepts in particular, the definition of technical volumes – the use of existing and dimensioning of new installation space – plays a central role in geometrically and functionally integrating the key components, which have been introduced in the preceding section.

Figure 6 shows the basic disposable installation spaces in the vehicle. These definitions are based on the assumption that the electric vehicle concepts that can be realised in the medium term are also characterised by a typical double-row seat layout and that the physical structure of the vehicle is derived from conventional concepts.

The installation spaces that are significantly relevant for the key electric components are configured in the floor assembly of the vehicle, thus in the physical "platform". This assumption is justified as several attributes of traditional vehicle architectures, for example safety features, dimensional layout, approaches for package, ergonomics and not least the fulfilment of technical and statutory requirements will remain valid in the field of alternative vehicle architectures in accordance with the purpose design approach. Even though further propositions such as OEM-specific requirements could have a direct influence on vehicle dimensions and consequently on the specification and morphology of installation spaces, several degrees of freedom emerge in the context of the integration of drivetrain and energy storage systems. Their potential is to be analysed. One essential aspect is the elimination of the combustion engine propulsion. Installation space can be configured underneath the passenger cell as well as in the front section and the rear of the vehicle according to design of the new key components.

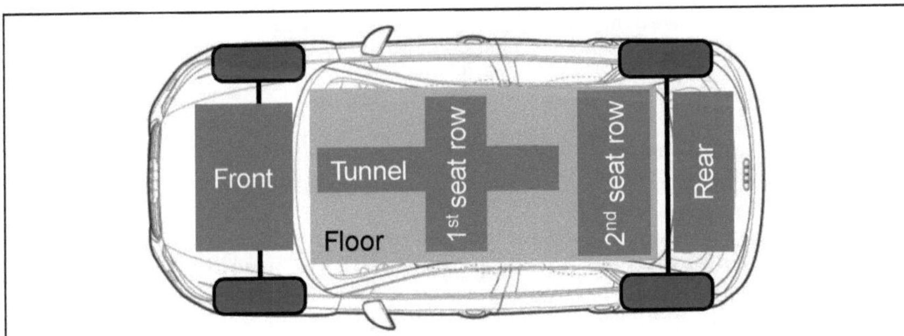

Figure 6: Available installation space in the physical vehicle platform

8.3 Design and analysis of electric vehicle concepts

To analyse the influences of electric components and their underlying technologies on the complete concept of electric vehicles, a reproducible and transparent process is essential. All relevant variables and impacts on the conceptual design in the early phase are to be taken into account. For this reason, an adequate compromise has to be defined between the depth and the scope of consideration.

8.3.1 Process model and examination scope

Figure 7 shows the process that has been developed against the background of the aforementioned challenges. It starts with a rough vehicle concept, which contains several items of basic information on the vehicle to be designed. These include the vehicle type and segment, related rough dimensions, drive layout, assumptions for package design as well as basic target values for characteristic concept variables such as driving performance, consumption, driving range and weight situation.

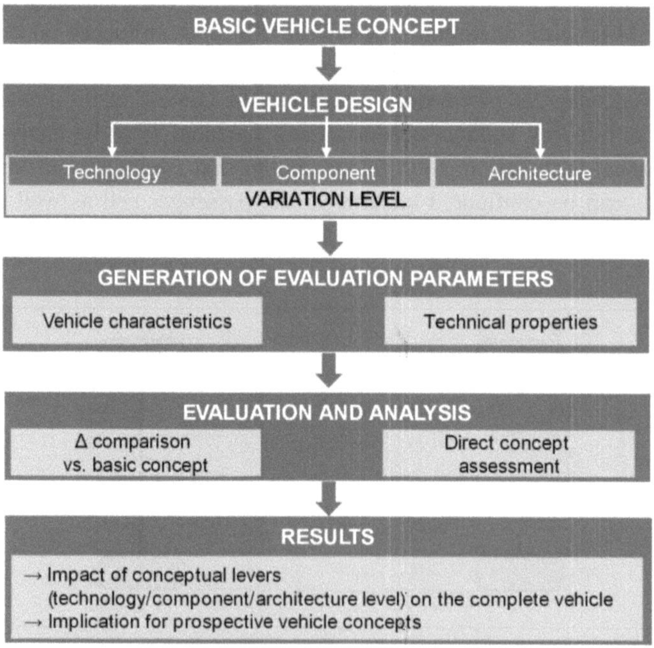

Figure 7: Process for concept design and analysis

Table 1: Conceptual levers on the variation levels technology, component and architecture

Technology	Component	Architecture
• **Electric energy storage** – Cell chemistry – Vol./grav. specific energy – Cost – Safety • **Electric drive** – Efficiency – Vol./grav. specific power – Cost • **Fuel cell** – Vol./grav. specific power – Cost – Efficiency – Vol./grav. storage density	• **Modularity vs. integration** – Single components/ integrated systems – Scalable components – Individual/standard parts • **Alternative modules** – Recharging system – Cell geometry	• **Drive topology** – Front/rear/all-wheel drive – Axle drive, single-wheel drive, in-wheel drive • **Storage topology** – Single/multipart storage system – Installation space • **Dimensional layout** – Segment – Basic architecture, flat/high/sandwich floor

Based on this, the design of the vehicle concept is operated: Energy storage and drive systems can be dimensioned by concept parameters both functionally and geometrically. In this context there are essential conceptual levers on the levels of technology, components and vehicle architecture that ensue in the course of design. Table 1 specifies some of these as an example.

The potential impacts of these levers on the next-higher investigation level and ultimately on the complete vehicle concept are to be analysed according to the procedure model shown in Figure 7. On the one hand, the subsequent generation of evaluation parameters results with reference to common vehicle concept parameters. On the other hand, concept assessment by means of technical characteristics can be performed, which cannot be directly allocated to established vehicle concept parameters or which are not defined in the rough concept (see Table 2).

The evaluation parameters generated by this can be used for assessment of different design results amongst each other, as well as for comparison with a reference vehicle, which is defined in the rough concept. Procedures based on the

Table 2: Assessable vehicle characteristics and technical properties

Vehicle characteristics	Technical properties
• **Performance/consumption** – Max. speed – Acceleration – Energy consumption – Range • **Weight situation** – Vehicle weight – Components weight – Axle load – Inertia – Payload • **Cost situation** – Purchase cost – Life cycle cost – Itemised cost • **Dimensional layout** – Basic dimensions – Dimension chains	• **Topology of installation space** – Shape of installation space – Use of installation space – Variance • **High-voltage architecture** – Number of parts – Interfaces – Wiring – Complexity, additional weight • **Qualitative impacts on technical domains** – Vehicle safety – Production/assembly – Service – Dimensional layout • **Suitability for platform/modular design** – Basic (high-voltage) architecture – Scalability of components/systems (function, geometry)

utility analysis with preceding definition of a target system provide a methodical support for evaluation and documentation. In principle however, other evaluation methods are also possible [6].

8.3.2 Description of scenarios and technology roadmaps – termination of the analysis focus

In general, the above-mentioned investigations and the possible input and output variables give a broad spectrum for analysis focuses and – in the course of an iterative vehicle design – a high number of possible alternative configurations that could be used to describe a basic vehicle concept with different variations of technology, component and architecture levels.

In order to perform a dedicated analysis, the investigation scopes and the variation ranges must first be defined. As previously described in Section 8.2.1,

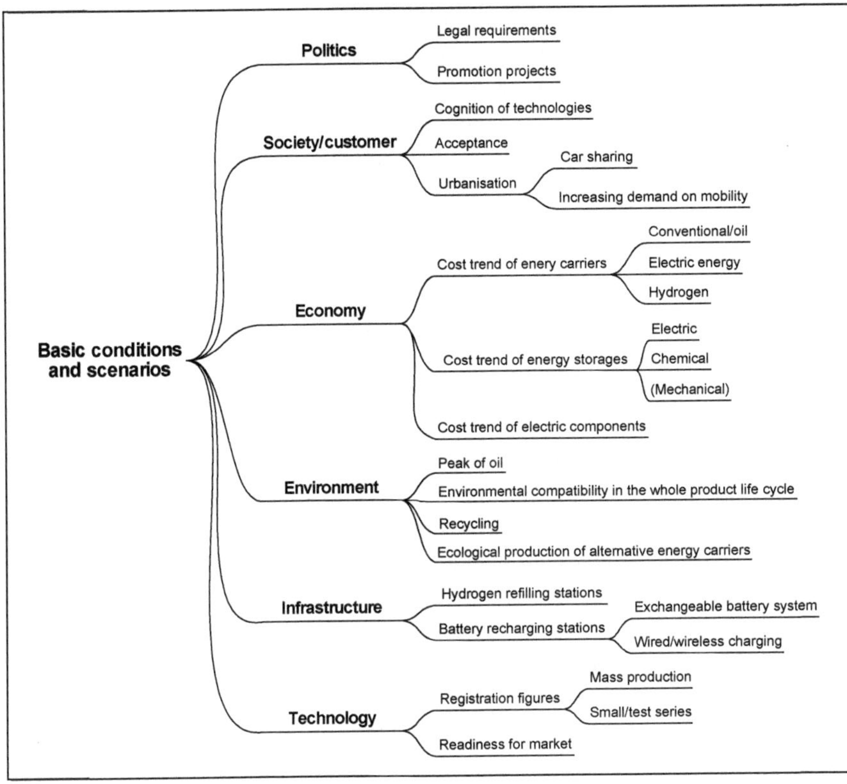

Figure 8: Relevant domains and specific factors for scenarios of electric vehicle concepts

the focus of the research approach presented here is on battery and fuel cell electric vehicles. Essential criteria for the chronologically classified relevance of these two types of vehicles in the concept development of an OEM can be derived by developing adequate scenarios.

The domains and aspects listed in Figure 8 influence possible scenarios for the comprehensive establishment of electric vehicle concepts. Here, the influence factors from politics, society, economy, environment, infrastructure and vehicle technology play a decisive role. These are accompanied by strategic decisions by the manufacturers to develop these concepts – as in this approach for BEV and FCEV – to series production vehicles. Examination assumptions, which are described below, are made on the basis of the analysed influences shown in Figure 8.

Fundamentally viewed, the technologies employed in BEV concepts are already developed. However, there is an enormous need for further development with respect to the integration of energy storage and drive system components into the vehicle. Making these technologies suitable for application in the automobile is particularly important. In the field of battery electric vehicles, a limited spectrum of vehicle concepts is already feasible. The first models are in series production and are available on the market.

The establishment of small city vehicles in the lower segments (small cars and compact cars, cf. [2]) with limited range for everyday commuter traffic is seen as a suitable launch scenario for BEV. Higher vehicle segments and applications will follow [7]. It is therefore appropriate to examine BEV concepts based on the current state-of-the-art in the field of electric vehicle technologies and to analyse possible development potential in the key components and their impacts on the concept design. Based on the configuration of BEV concepts for small and compact vehicle segments, it must be examined to what extend battery-powered vehicles are feasible and reasonable in larger vehicle classes.

Fuel cell electric vehicles are seen as a possible solution with respect to zero-emission long-range mobility for everyday use. Some carmakers have been working on technologies for the necessary energy storage and converters. The proton exchange membrane fuel cell (PEM), which uses hydrogen and atmospheric oxygen as reaction media, is regarded as the most promising concept. Despite all efforts on the part of the manufacturers, it is still a long way from being introduced into fuel-cell-powered mass-produced vehicles. Firstly, the actual technologies and components – i.e. fuel cells and hydrogen storage systems – have to be developed to series production maturity. Secondly, there are further influencing factors that also determine the introduction scenario and therefore the chronological classification of FCEV. Based on the high cost of fuel cell concepts, it can be assumed that the establishment of series production vehicles will start in the upper vehicle classes. Against this background, it is rational to analyse FCEV concepts for fully-fledged long-distance vehicles in the middle and upper class segments and to examine the adaptation potential for other vehicle segments.

The research scenario described above is graphically summarised in Figure 9.

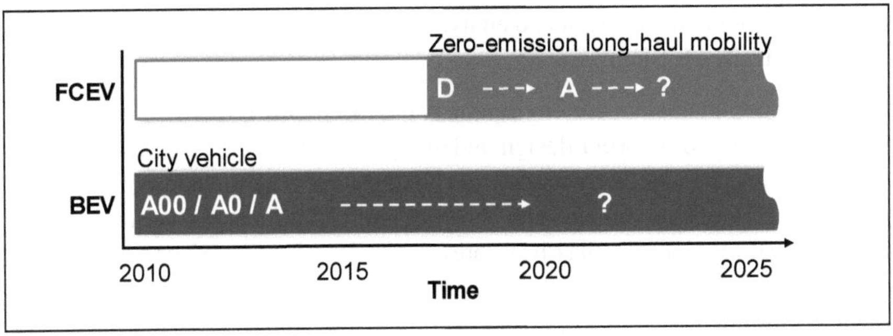

Figure 9: Assumption of a chronological scenario of electric vehicle concepts

8.3.3 Tool chain for computer-aided concept development and analysis

In order to analyse the described operations in accordance with the process explained in Section 8.3.1, efficient support by suitable methods and tools is required. For this reason, tools for functional and geometrical concept design and analysis have been developed. This has created a computer-based environment that describes all relevant technical interactions in the vehicle concept.

The individual subsystems of the tool chain were implemented on the basis of MS Excel, MATLAB/Simulink and CATIA V5. This satisfies the requirement explained in Chapter 1 that both functional and geometric concerns are taken into account in the configuration of components and of the complete vehicle.

Figure 10 illustrates the basic architecture of the described method chain. The mathematical design and analysis is performed using a tool implemented in MATLAB/Simulink. MS Excel represents a consistent data base that also acts as the interface to the CAD software CATIA V5. The geometric concept development is performed here using a Digital Mock-Up (DMU).

Figure 10: Framework of a methods environment for functional and geometrical design and analysis

8.3.3.1 Mathematical design and analysis environment based on MATLAB/Simulink

The developed mathematical environment based on MS Excel and MATLAB/Simulink is used for concept design and analysis with the aid of component and vehicle parameters linked by analytical relationships. The methodology of linking parameters of the vehicle with those of individual concept variables using mathematical relationships has been previously described in [8] and [9]. Newer approaches are presented for example in [10] and [11], where the variation of selected concept variables is pursued to optimise characteristics of rough concepts for battery electric vehicles. The extent of analyses which are necessary in accordance with the details given in Chapter 8.1 and in Section 8.3.1 create further functional requirements on the configuration tool:

- Configuration of key components of BEV and FCEV concepts with suitable analytical models
- Description of the interactions between technology, component and complete vehicle parameters
- Enabling of multidirectional design and parameter variations
- Performance of complete vehicle calculations and simulations

In addition, the following requests exist with respect to the model architecture:

- Modular configuration with a view to expandability and exchangeability of individual operations
- Active user integration, possibility of manual intervention in component and vehicle design
- No obstruction of degrees of freedom as a result of excessive automation/optimisation
- Definition of interfaces to other dimensioning domains (MS Excel, CATIA V5)

Thus, the developed design and analysis environment permits a consistent rough draft of key electric components

- Electric drive (machine, gearbox)
- Electric energy storage (traction battery)

- Fuel cell stack and
- Hydrogen storage
- High-voltage supply

as well as the performance of calculation and simulation operations on the complete vehicle level to determine the variables listed in Table 2.

As additional components of the high-voltage system, power inverters, high-voltage distributors, DC/DC converters, recharging devices, high-voltage wiring as well as PTC heating and air conditioning compressor are taken into account to describe their influence on the complete vehicle, and especially on the topology of the high-voltage system. The principal workflow of the design and analysis environment is shown in Figure 11.

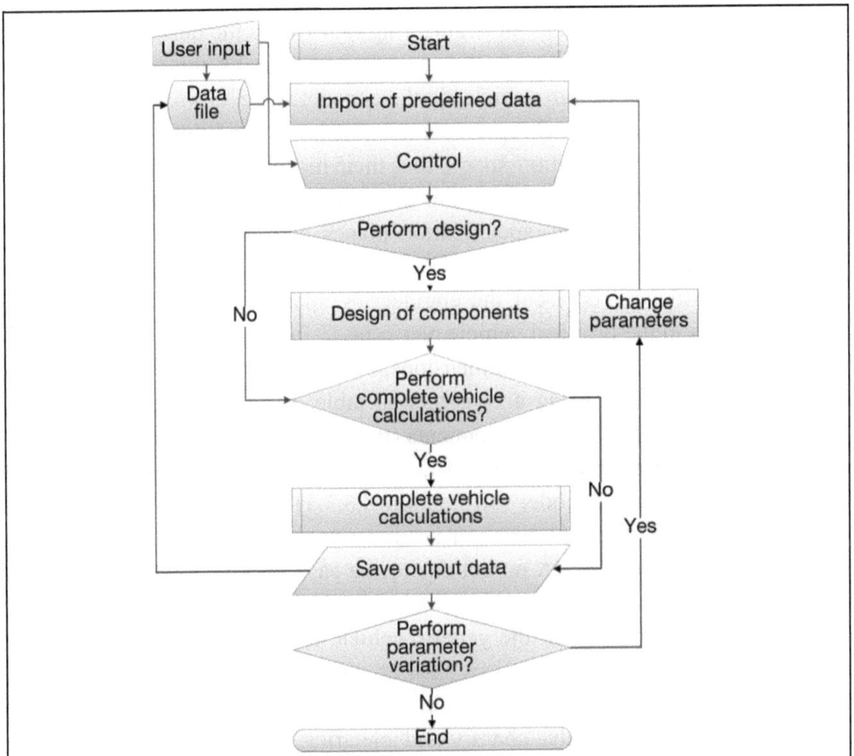

Figure 11: Simplified flow-chart of the MATLAB/Simulink-based calculation programme

Here it is essential that sub-functions can be activated and deactivated separately. This allows, for example, the configuration of the electric drive system using either the implemented algorithms or pre-specified data. The same degrees of freedom also exist for the other component configurations as well as for calculations and simulations for determining the complete vehicle variables. According to Figure 10, input and output of data for components and the complete vehicle are provided by an MS Excel structure. The individual setups for the required calculation operations are made via user dialogues in the MATLAB main programme.

8.3.3.2 Generic vehicle concept model (concept DMU) based on CATIA V5

As a method of virtual product development, the Digital Mock-Up is of central importance within the context of the project presented here. The virtual concept model is actively integrated as an engineering tool within the design and analysis process. This is done according to the procedure used in the automotive industry in order to technically design vehicles in the concept stage.

The vehicle concept model represents the geometric conditions in the vehicle: dimensional layout, installation spaces and their morphology, vehicle structures as well as components that are essential for concept design and analysis are implemented here. In addition, specification planes and design premises such as ground clearance surfaces, released package spaces and structural patterns can be illustrated in order to represent the engineering space that is relevant for geometric analyses in the physical vehicle platform.

Explicitly definable vehicle dimensions, including standardised dimensions [12, 13], are controlled using a construction table. Installation space specifications for concept-determining components (traction battery, drive system, hydrogen storage) are applied in the same way. Adequate algorithms implemented in the MATLAB/Simulink model (cf. Section 8.3.3.1) can perform their package configurations. Other parameters can be directly manipulated via the structure tree of the parametric CATIA model due to the easy accessibility and the visual support.

One essential relevance of the generic vehicle concept model can be seen in the consideration of design premises such as those resulting from specifications of the technical domains. Specific requirements can be derived from the areas of the drivetrain, vehicle body, chassis and electrics, which are responsible for several components, as well as from multidisciplinary domains such as vehicle

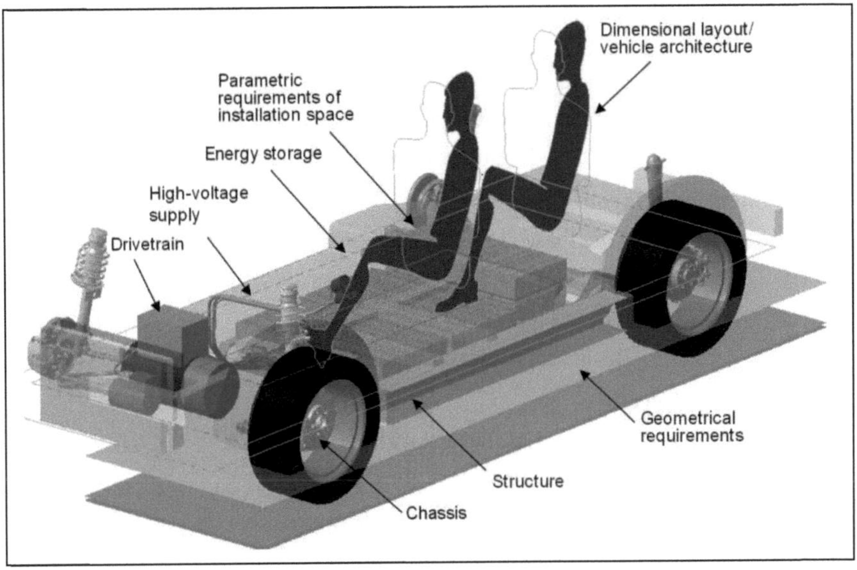

Figure 12: Generic vehicle concept model (CATIA V5)

safety, production/assembly and service/after-sales. Partly, these are directly reflected in the functional and geometrical vehicle design. A structuring and analysis of these requirements for BEV concepts shows that much of this affects the arrangement and specification of the traction battery and the arrangement of the drivetrain components.

As an example, Figure 12 shows the reference vehicle for a BEV concept in the A0 segment, which is explained in Chapter 4 comprising the implemented scopes being implemented in the generic DMU in a simplified form.

8.4 Application of the analysis process: Potential of electric energy storages for prospective BEV concepts

8.4.1 Reference vehicle concept

The following chapter shows the application of the procedure by means of the BEV scenario presented in Section 3.2 for selected conceptual levers (cf. Table 1) in the domain of the traction battery system. The reference vehicle concept illustrated in Figure 12 constitutes the starting point, which is characterised by the conceptual parameters listed in Table 3.

Table 3: Concept characteristics of the BEV A0 reference concept

Vehicle type	
Level of electrification	BEV
Segment	A0, city vehicle
Raw dimensional layout	
Wheelbase (L101)	2540 mm
Vehicle length (L103)	3890 mm
SGRP to ground (H5-1/-2)	626 / 644 mm
Basic architecture	Sandwich floor
Drivetrain architecture	
Type	Axle drive front
Electric power (peak)	85 kW
Gear	Single-speed, coaxial
Energy storage	
Type	Li-ion ($LiCoO_2$)
Cell standard	Prismatic, PHEV-2 [14]
Configuration	3p100s
Cell capacity	28 Ah (corr. ~145 Wh/kg)
Nominal energy	31 kWh
Installation space	2nd seat row, sandwich floor
Driving performance	
Max. speed	160 km/h
Acceleration 0-60 km/h	5.5 sec.
Range/consumption NEDC	
Range	200 km
Consumption	13 kWh/100 km

8.4.2 *Variation of electric parameters of battery cells*

On the levels of component technology, the gravimetric specific energy can by identified as a concept-influencing lever. On the basis of a roadmap analysis for lithium-ion battery technologies, the scenario shown in Figure 13 becomes apparent. According to optimistic estimates, essential improvements with a production-ready state of development of approximately 250 Wh/kg (cell level) can be achieved within the next eight years. Predestined follow-up technologies such as lithium-sulphur and lithium-air systems are not considered in this respect.

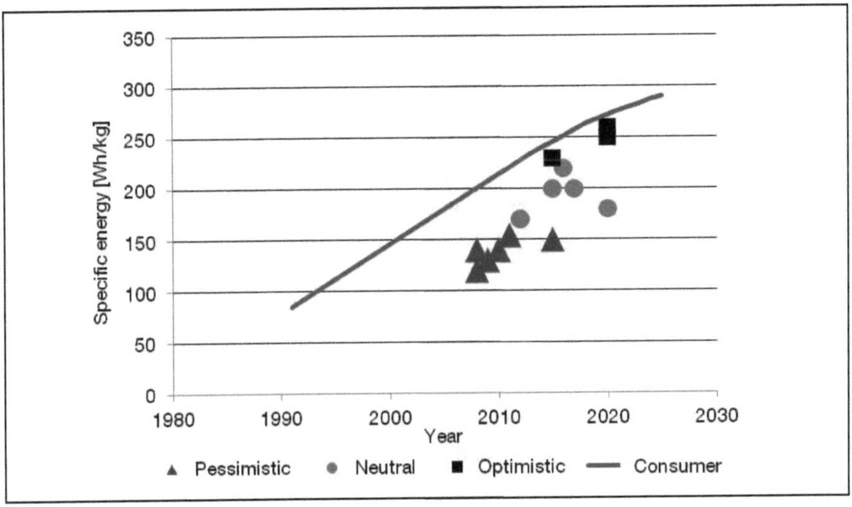

Figure 13: Roadmap for specific energy of Li-ion battery cells for automotive application

Applied to the battery cell of type PHEV-2 [14], which is used in the reference vehicle, this results in a cell capacity of up to 48 Ah. If the system configuration remains constant, the energy content could be enhanced by more than 70 percent. Corresponding high driving ranges of more than 350 kilometres (NEDC) can be realised in this vehicle concept. However due to the use scenario as a city car, the unrestricted increase of the driving range is not reasonable.

Instead, the accessible potentials can be used for optimisation of the system configuration and thus of the complete energy storage layout. In this context, Figure 14 illustrates the existing restrictions in system layout of battery assemblies. These are composed of single cells, which are connected in series and in

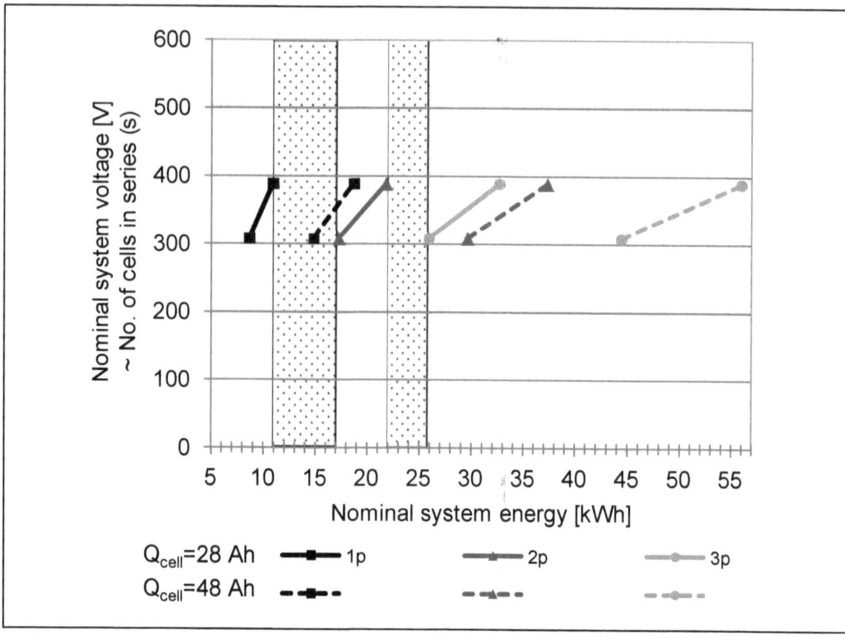

Figure 14: Design restrictions for battery system configuration

parallel (full lines). The minimum and maximum number of cells in series is restricted by predefined minimum and maximum values for system voltage. It is therefore clear that certain energy contents cannot be configured at all or only with a dedicated number of parallel strings. These dimensioning gaps also appear in the geometric concept layout, because disposable installation space cannot be varied to the same degree and in the same step size as the system configuration of the traction battery. Inevitable steps in the number of cells and spatial needs for required energy content and system power cannot be configured at all. Enhancement of cell capacity as well as increase of system voltage level represent an essential lever for the medium-term optimisation of the design of traction battery systems.

Taking target values for performance and driving range according to Table 3 into account, a reconfiguration of the battery system allows the identification of the primary impacts on the energy storage and the overall vehicle: In consequence of increased cell capacity, the configuration of energy content due to a driving range of about 200 kilometres is enabled by a double-string battery system (Figure 14, dashed lines).

Table 4: Excerpt of parameters of components and complete vehicle for varied battery cell capacity

Cell capacity	**28 Ah**	**36 Ah**	**48 Ah**
System configuration	3p100s	3p84s	2p86s
Nominal energy	31 kWh	33 kWh	30 kWh
Installation space			
1st seat row	✓	✓	✓
2nd seat row	✓	✓	✓
Sandwich, front	✓	✓	O
Sandwich, back	✓	O	O
Dimensions			
Wheelbase, overhang		Const.	
Vehicle height (H100)	1539 mm	1521 mm	1521 mm
Axle load front	53.6 %	54.2 %	55.1 %

Table 4 contrasts some resulting parameters for components and the complete vehicle with the reference vehicle. The potential of enhanced cell capacity in consideration of the system dimensions is evident. The resulting reduced demand on cells allows a more flexible use of installation space and dimensional layout. Evaluation of weight distribution ensues that the positioning of the battery system can be used for systematic influence or even adjustment of axle-load distribution. Furthermore, the seating position in the rear compartment can be lowered in order to reduce the vehicle height. These steps are reflected in a more sportive design as well as decreased energy consumption by lower aerodynamic drag.

Figure 15 shows the impacts on several parameters of the battery system expressed in percent. The resulting modifications in installation space, possible package optimisation and cost-reducing degression are considered.

The degrees of freedom for a more flexible spatial morphology shown in Table 4 offer the feasibility of design of geometrically scalable battery systems in a further step. These can be dimensioned for different derivatives of a basic vehicle (e.g. flat/high floor models). In addition, synergies covering the complete model range can be created by using the same energy storage system for different vehicle segments, thanks to its flexible arrangement in the longitudinal plane of the vehicle (cf. Figure 16).

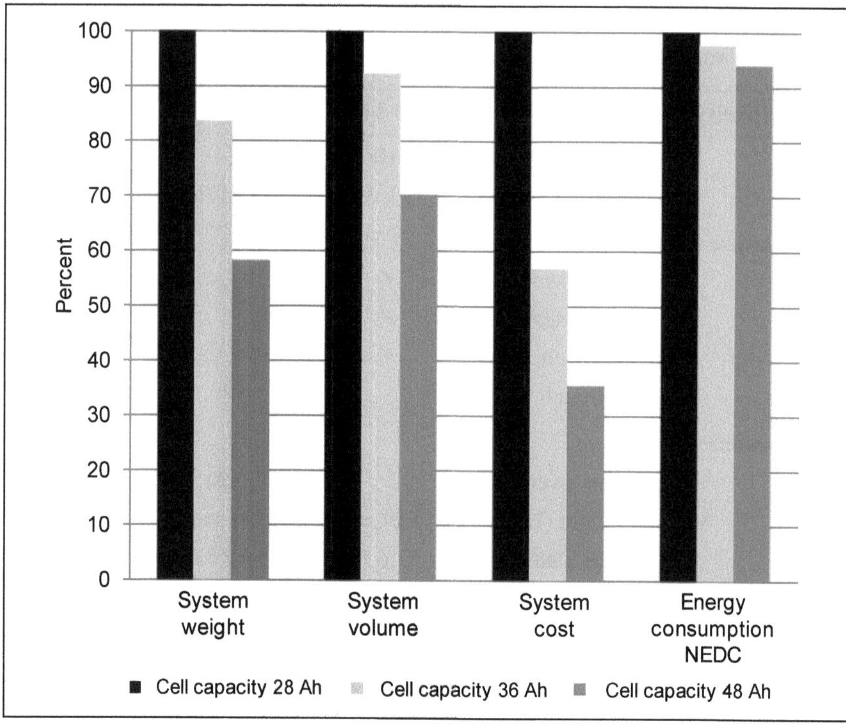

Figure 15: Parameters of battery system for varied battery cell capacity

Thus, the prerequisites are created for the establishment of a modular design principle covering all model ranges and segments. According to the state of technology that was shown in the reference vehicle concept, such a building block concept would not be feasible. The ratio of available to required installation space would not enable a flexible integration of the energy storage.

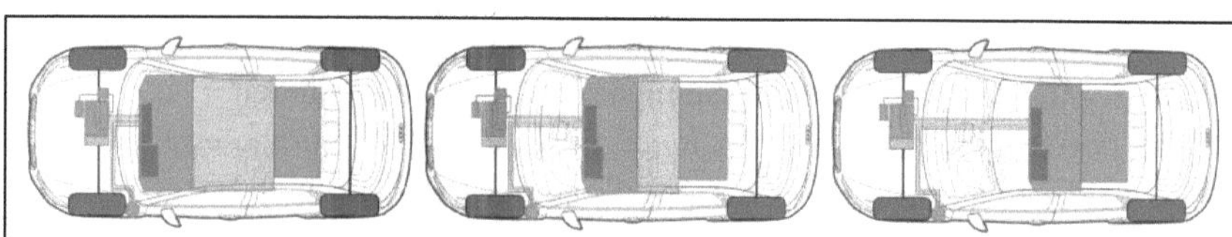

Figure 16: Scalable arrangement of the battery system

8.4.3 Variation of installation space for the battery system – potential of enhanced safety features

Influences from design of traction battery on the high-voltage and vehicle architecture are explained by means of an additional example. In addition to the improvement in the energy and performance figures of battery cells, manufacturers are pursuing a significant increase in safety features. Electric short-circuit faults and the resulting fire hazard, as well as leakage of hazardous substances, are amongst the most significant risks associated with an accident with an electric vehicle. By an adequate arrangement of the battery system in the vehicle, it is possible to avoid the initiation of these hazard mechanisms in the case of a crash. Beyond that, suitable structures can be provided in the vehicle as well as in the battery system itself in order to provide protection for the battery cells. However, these measures serve only to reduce the incidence rate of damage and the minimisation of the damage consequences. The real lever for improving the safety features of traction battery systems lies in intrinsically safe cell chemistry and in the use of adequate separators and electrolytes.

Crash analyses show that the arrangement of the battery system between the wheel axles and underneath the passenger cell as well as if possible away from the exterior structures of the vehicle must be the preferred means of protecting battery cells against mechanical stresses. Vehicle zones with various safety levels for installation of high-voltage components are shown in Figure 17 according to [16] and [17].

Energy storage and drive components should therefore preferably be integrated in the vehicle sections designated as zone A and zone B. Provided that the battery technology used is intrinsically safe – on the basis of mixed material

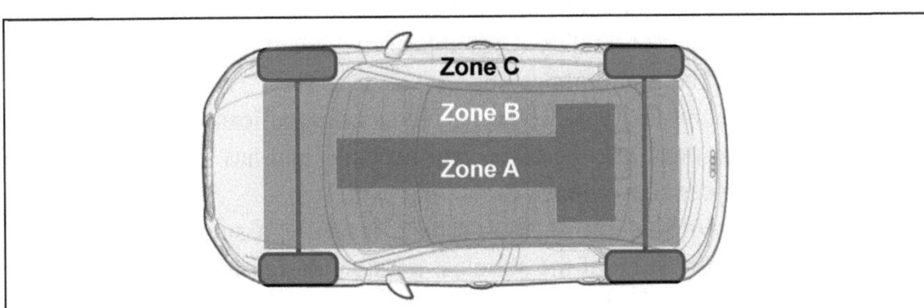

Figure 17: Packaging restrictions for high-voltage components according to [16] and [17]

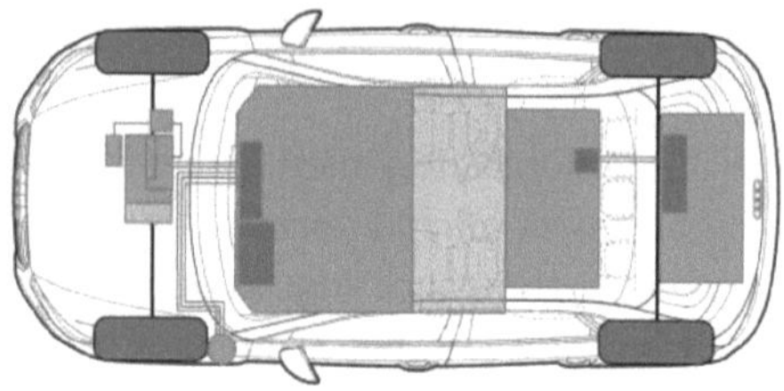

Figure 18: Alternative topology with battery at rear

cathodes for instance – it is possible to use these installation spaces, which are classified as inappropriate according to the definition shown in Figure 17 above, without additional structural measures. This could be the rear of the vehicle in zone C. The utilisation of this additional degree of freedom enables an alternative topology of the energy storage, which is divided into two separate parts in a corresponding redesign (cf. Figure 18).

Table 5 gives a comparison of the changes in essential parameters. As documented in Table 4, there arises a potential for the reduction of the vehicle height by lowering of the seating position in the rear compartment. However, the modifications in the topology have a negative effect on the axle-load distribution, which is outside the usual range for front-wheel-drive vehicles [19] and cannot be corrected sufficiently by modification of the battery position in the underfloor space.

Impacts on the architecture of the high-voltage system are shown in Figure 19. As can be seen, the improvement of additional installation space is connected with substantial additional technical costs, which causes a higher system complexity of the high-voltage supply as well as an increased weight and cost of the complete system. Due to an additional battery subsystem, necessary system parts such as electric contacts and safety fuses appear redundant. Furthermore, additional connections by the high-voltage supply are necessary. The number of interfaces, the system complexity and cost are increased significantly. The increased manufacturing cost must also be taken into account, and has not yet been considered in the analyses shown here.

As a conclusion of this examination aspect, it can be assumed that basically one coherent battery system of the simplest possible design is the ideal solution. By use of the potential of safe battery cells for additional installation space in the rear of the vehicle, several severe concept weaknesses cancel out the initially obvious advantages for the complete vehicle concept.

Table 5: Excerpt of parameters for alternative topology of the battery system

Storage architecture	**Reference concept**	**Alternative concept with battery at rear**
System configuration	3p100s	3p100s
Nominal energy	31 kWh	31 kWh
Installation space		
1st seat row	✓	✓
2nd seat row	✓	✓
Sandwich, front	✓	✓
Sandwich, back	✓	O
Rear	O	✓
Dimensions		
Wheelbase, overhang	Const.	Const.
Vehicle height (H100)	1539 mm	1521 mm
Axle load front	53.6 %	50.25 %

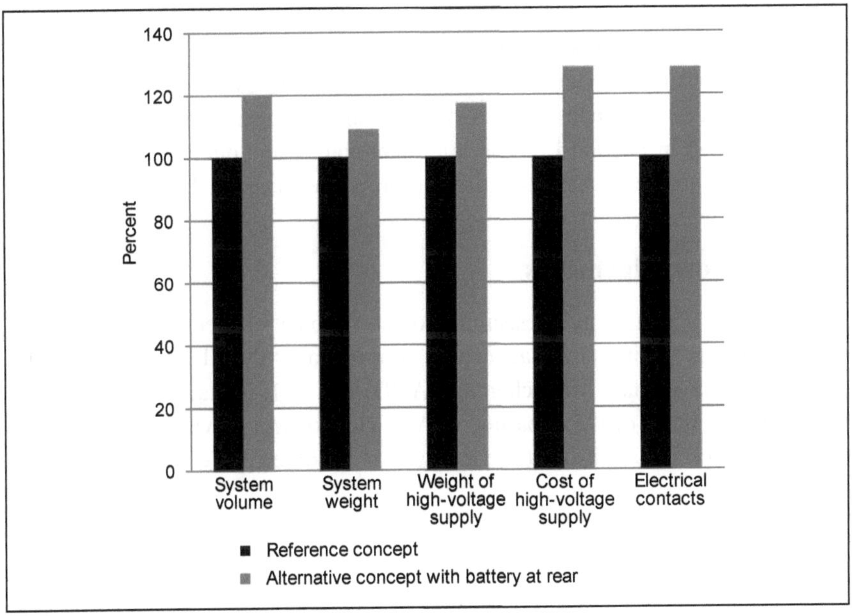

Figure 19: System parameters for varied battery topology

8.5 Conclusion and outlook

This paper shows the challenges in design of prospective electric vehicle concepts. A methodical process model for design and analysis of interdependencies between technology, components and complete vehicle has been introduced. In support of the necessary analyses, a tool chain based on MATLAB/Simulink, MS Excel and CATIA V5 has been developed for this purpose. As opposed to approaches that pursue optimisation of rough vehicle concepts, the project described in this paper systematically examines the specific influences of electrification on the vehicle architecture and high-voltage topology. The exemplary application of this procedure to the importance of possible levers in the field of the traction battery system for BEV concepts showed significant potentials for prospective development of battery electric vehicles. The analysis of the results showed that a fundamental understanding of functional and geometrical interactions is essential for the effective use of technological potentials for the design of prospective vehicle concepts. Additional technical levers on the levels of vehicle architecture, key electric components and their underlying technologies must be analysed and evaluated. Over that, the examination illustrated the importance of regarding both customer relevant vehicle characteristics and technical issues in the design of electric vehicle architectures.

The defined process model and the methods it contains serve as the basis for a technical assessment system. Thus, the specific influences of electrification during concept development can be evaluated and documented with regard to the issues described in Section 1.3. Essential quantitative and qualitative parameters can be derived from the characteristics and properties listed in Table 2.

8.6 Acknowledgements

This paper is a part of the research project “Impact of electrification on the vehicle concept” within the academic cooperation “INI.TUM” between the Technische Universitaet Muenchen and AUDI AG in Ingolstadt. Special thanks go to the colleagues of the department “Vehicle Concept Development” as well as to colleagues and students at the Institute of Automotive Technology for their support.

8.7 References

[1] Wallentowitz, H.; Freialdenhoven A.; Olschewski I. “Strategien zur Elektrifizierung des Antriebsstrangs”, Vieweg+Teubner, Wiesbaden 2010.

[2] Braess, H.; Seiffert, U. “Vieweg Handbuch Kraftfahrzeugtechnik”, 4. Auflage, Vieweg/GWV, Wiesbaden 2005.

[3] Hofmann, P. “Hybridfahrzeuge”, Springer, Wien 2010.

[4] Wallentowitz, H.; Reif, K. “Handbuch Kraftfahrzeugelektronik”, 1. Auflage, Vieweg/GWV, Wiesbaden 2006.

[5] Lindemann, U. “Methodische Entwicklung technischer Produkte”, Springer, Berlin 2005.

[6] Breiing, A.; Knosala, R. “Bewerten technischer Systeme”, Springer, Berlin 1997.

[7] Bundesministerium für Verkehr, Bau und Stadtentwicklung (BMVBS) “Nationaler Entwicklungsplan Elektromobilität”, Berlin 2009.

[8] Heinke, O. “Fahrzeugauslegung mit Hilfe von Eigenschaftsparametern – Möglichkeit oder Utopie?”, VDI-Verlag GmbH, Düsseldorf 1994.

[9] Braess, H.; Stricker, R.; Baldauf, H. “Methodik und Anwendung eines parametrischen Fahrzeugauslegungsmodells”, Automobilindustrie 5/85, p. 627-637.

[10] Wiedemann, E.; Meurle J.; Lienkamp, M. “Vom Kundenwunsch zum elektrifizierten Fahrzeugkonzept”, 2. Automobiltechnisches Kolloquium, Garching 2011.

[11] Kuchenbuch, K.; Stieg, J.; Vietor, T. “Individual concepts for electric vehicles: Interaction between battery package and vehicle concept”, 11th International Stuttgart Symposium Automotive and Engine Technology, Stuttgart 2011.

[12] Astier, M. “Global Cars Manufacturers Information Exchange Group: Package Drawing Exchanges”, Renault s.a.s., Boulogne-Billancourt (F) 2008.

[13] DIN 70020-1 “Straßenfahrzeuge – Kraftfahrzeugbau – Begriffe von Abmessungen”, Beuth, Berlin 1993.

[14] DIN SPEC 91252 "Elektrische Straßenfahrzeuge – Batteriesysteme – Abmessungen für Lithium-Ionen-Zellen", Beuth, Berlin 2011.

[15] Hollmotz, L.; Hackmann, M. "Sicherheit von Lithium-Ionen-Batterien im Kraftfahrzeug – Stand der Technik, Risiken, Zielkonflikte, Testverfahren und Bewertungsmethoden", VDI-Berichte Nr. 2144, Fahrzeugsicherheit – Fokus Elektromobilität, Berlin 2011.

[16] Schaub, N. "Erforderliche Maßnahmen zur Prüfung (z.B. Crash) alternativ angetriebener Gesamtfahrzeuge und deren Komponenten", VDI-Berichte Nr. 2144, Fahrzeugsicherheit – Fokus Elektromobilität, Berlin 2011.

[17] Barenyi, B. Patentnummer 854157, "Kraftfahrzeug, insbesondere zur Befoerderung von Personen", 28.08.1952.

[18] Ketterer, B.; Karl, U.; Möst, D.; Ulrich, S. "Lithium-Ionen Batterien – Stand der Technik und Anwendungspotenzial bei Hybrid-, Plug-In Hybrid- und Elektrofahrzeugen", Forschungszentrum Karlsruhe GmbH, Karlsruhe 2009.

[19] Heißing, B.; Ersoy, M.; Gies, S. "Fahrwerkhandbuch", 3. Auflage, Vieweg + Teubner, Wiesbaden 2011.

9 Vehicle Dynamics per Software – Potentials of an Electric Single Wheel Drive

Dr. Alfred Pruckner, Stephan Kaspar, Dr. Ralf Stroph, Dr. Christoph Grote

Abstract: This paper outlines the advantages of drive train architectures which allow free distribution of wheel torques on the rear wheel axle, so-called torque vectoring, with respect to driving dynamics. Electric cars open up new possibilities with regard to vehicle architectures. Utilization of the new degrees of freedom might lead to architectures with rear wheel drive and high rear axle loads. For conventional cars without counteractive measures, e.g. mixed tires, this would inevitably lead to inappropriate driving behavior. It is shown that such boundaries are no longer valid for electric cars with high rear axle load in combination with rear wheel drive and torque vectoring.

Keywords: torque vectoring, electro mobility, driving dynamics, rear wheel drive.

9.1 Change in vehicle architecture

Established car manufacturers have to make a fundamental decision at the beginning of the development of an electric vehicle: Either using an existing, conventional combustion engine driven car and convert it to an electric one by changing the drive train and its appendant components, so-called conversion design; or newly develop a completely independent platform combining all advantages and peculiarities of electro mobility that can be taken into account, which is called purpose design.

Often, manufacturers want to avoid the cost and development risks of new platform architectures, so they decide to use conventional vehicles as a basis, for example BMW for their ActiveE (cp. Figure 1). Mainly motivated by cost, effort and uncertainty reasons, the drive train concept is rather adapted than fundamentally changed, which means one driven axle by one engine.

If the decision is made in favor of a whole new architecture, special boundary conditions caused by the electric drive train and its components can be taken into

Figure 1: BMW ActiveE [1]

consideration. New degrees of freedom allow a more flexible arrangement of components [2]. Installation space requirements are one of the key factors. Indeed, there is available space for new components as conventional parts like the combustion engine or the exhaust gas system are no longer required. Besides the electric motors and the corresponding parts such as power electronics, the batteries which are large and heavy (around 10 kg/kWh) [3] have to be fitted into the vehicle. Even though 52% of costumers could think of buying an electric vehicle in the next five years [4], to establish electro mobility, vehicles have to be suitable for daily use [5]. This requirement leads foremost to a wide cruising range (which means large batteries) and no adverse effect on interior space. Furthermore, batteries must be shielded as well as possible to prevent hazardous situations in the event of a crash. Hence, placing the batteries at a place which is secure and not reducing passenger space is one of the key tasks for electric vehicles.

One problem-solving approach is the use of motors, attached closely to the wheel, one for each driven wheel [6] (cp. Figure 2, [7]).

As the combustion engine, the exhaustion gas system as well as the gearbox and the rear-axle differential can be omitted, enough space between the engines is released for the battery. That way, no installation space is needlessly wasted and the whole electric drive train with all needed components can be fitted into the car without limiting the passenger compartment or luggage space.

Figure 2: Optimized utilization of installation space by single wheel drive and compact battery package

Placing the engines, gears, power electronics and batteries near the rear axle results in large rear axle loads. For a conventional car, this could lead to an inacceptable driving behavior which might lead to stability problems, as shown in the next section.

9.2 Heavy rear axle loads and its effect on vehicle stability

As shown in the previous section, using the flexibilities and advantages of electric drives could lead to architectures with a rear wheel drive in combination with a high axle load. To comprehend the effects caused by this, one has to understand the behavior of the tires, as shown in Figure 3.

It can clearly be seen that the behavior is not linear but degressive: meaning, doubling the vertical load does not double the maximal transmittable horizontal tire forces. It can also be seen that, generally speaking, the degression is more remarkable for narrow than for wide base tires.

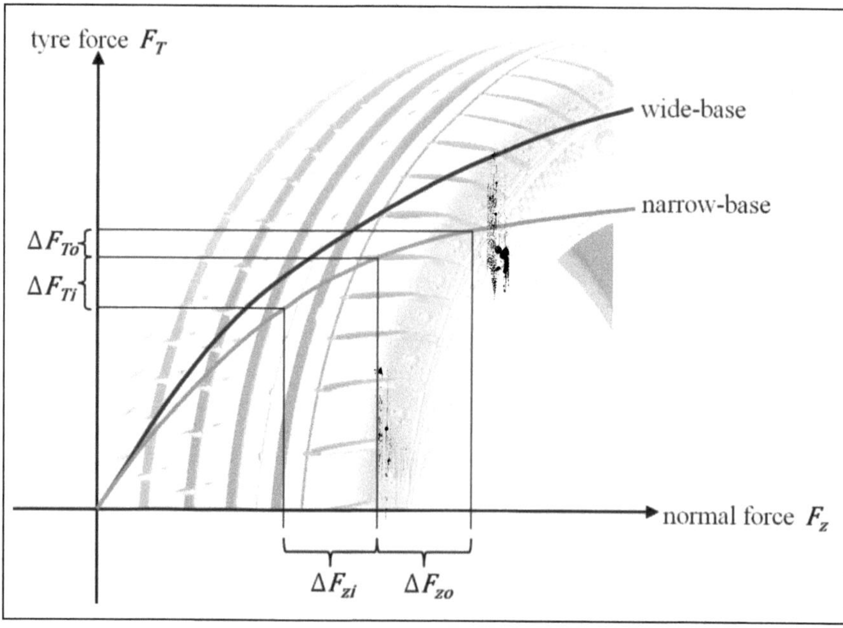

Figure 3: Degressive tire force potential

This characteristic affects the driving behavior, especially while cornering at lateral dynamic limits. Figure 4 shows a car from top view while cornering without longitudinal acceleration. The circles around the tires symbolize the traction limits of each wheel: Longitudinal and lateral forces can be chosen freely, but its vectorial sum cannot exceed the circle [8]. If the rim of the circle is reached, lateral force is reduced as longitudinal forces are the constraining forces.

If a car is driven around a corner at its lateral dynamics limits without longitudinal acceleration, the whole tire force potential can be used as lateral force. During steady state cornering, the yaw rate is constant, resulting in zero yaw moment. Figure 4 shows lateral forces' influences on the yawing of the car: While lateral forces of the front tires lead to an inward-turning yaw moment, the one of the rear axle is outward-turning. This means that for steady-state cornering yaw moments of front- and rear axle should be eliminating each other. Because of the previously shown degressive behavior, the position of center of gravity has an impact on the resulting yaw moment while cornering.

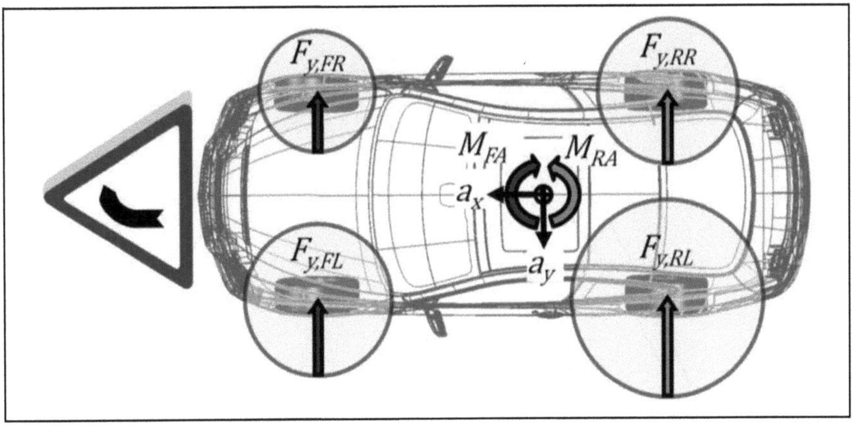

Figure 4: Tire forces while right cornering without longitudinal forces

To calculate the resulting tire forces, the normal forces $F_{z,ij}$ for each tire must be known for the actual driving situation according to longitudinal and lateral accelerations a_x and a_y, which can be estimated by [9]:

$$F_{z,FL} = m[(1-\eta_{Ha})g - \eta_{Rsp}a_x]\left[\frac{1}{2} - \frac{a_y h}{b_v g}\right] \tag{1}$$

$$F_{z,FR} = m[(1-\eta_{Ha})g - \eta_{Rsp}a_x]\left[\frac{1}{2} + \frac{a_y h}{b_v g}\right] \tag{2}$$

$$F_{z,RL} = m[\eta_{Ha}g + \eta_{Rsp}a_x]\left[\frac{1}{2} - \frac{a_y h}{b_v g}\right] \tag{3}$$

$$F_{z,RR} = m[\eta_{Ha}g + \eta_{Rsp}a_x]\left[\frac{1}{2} + \frac{a_y h}{b_v g}\right] \tag{4}$$

where i indicate the front/rear wheels (F, R) and j left/right (L, R). η_{Ha} is the rear axle load, g the acceleration of gravity, η_{Rsp} the wheelbase to the height of the center of gravity ratio, h the height of the center of gravity and b_v the track width. Influences due to road incline have not been taken into account. With knowledge of the actual normal tire force, maximum horizontal tire force potentials have been evaluated by means of a look-up-table. Finally, potential yaw moments of both axles are calculated dependent on the position of the center of gravity and compared with each other. The important factor for a yaw

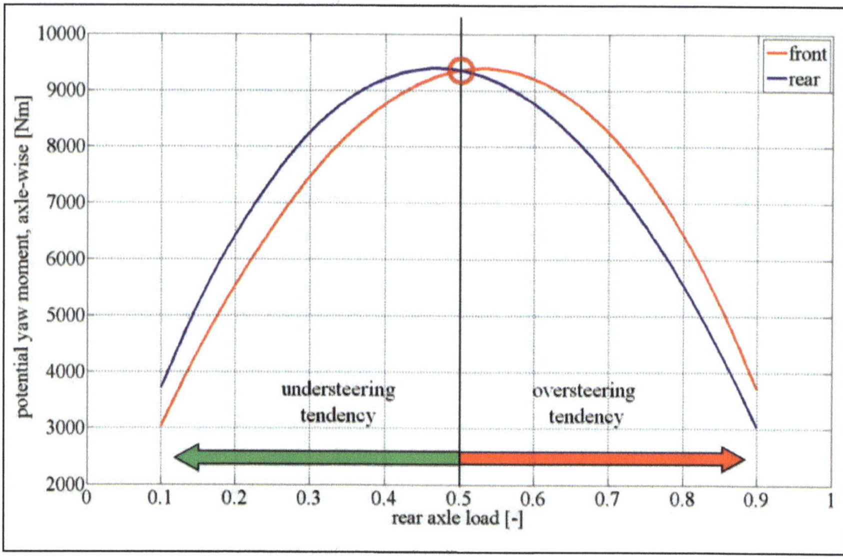

Figure 5: axle-wise yaw moments to assume the tendency of over- and understeering dependent on axle load distribution

movement is not the absolute values but their relation to each other. For better comparability the magnitude of the yaw moment was observed. The results are shown in Figure 5: the blue line indicates the maximum potential yaw moment of the rear axle, the one of the front axle corresponds to the red line. Neglecting influences of spring-damping-systems and anti-roll bar, a car with 50% rear axle load is perfectly balanced. With more front-heavy weight distribution, the influence of the rear axle increases compared to the front axle, which results in an understeering tendency. The car behaves safe, but not agile. For heavier rear axle loads the car tends towards oversteering (the magnitude of the potential yaw moment of the front axle exceeds the magnitude of the rear axle). Since it is easier for most drivers to handle under- than oversteering, the latter is the more critical tendency and has to be avoided.

9.3 Potential analysis of torque vectoring with respect to yaw moment

One of the key tasks to establish electro mobility is to expand the possible driving range of electric vehicles. A well recognized way is to use the electric engines as generators to restore energy while decelerating. For the investigated vehicle architecture (rear wheel drive) this means braking as much as possible by only using the rear wheels, which might lead to stability problems.

Again, a car cornering at lateral dynamic limits is examined, but this time with longitudinal forces applied to the rear wheels. It is assumed that the vehicle uses the whole tire force potential, which means application of longitudinal force for one tire involves a loss of lateral force, along the rim of the tire force circle as shown before. This would lead to even more oversteering. If it is possible to freely distribute drive- and braking torques, a yaw moment can be applied to stabilize the vehicle [10]. Figure 6 shows the results for different rear axle loads and longitudinal accelerations.

Based on calculated normal forces applied on each tire for two different rear axle loads and several driving situations (lateral acceleration at physical limits

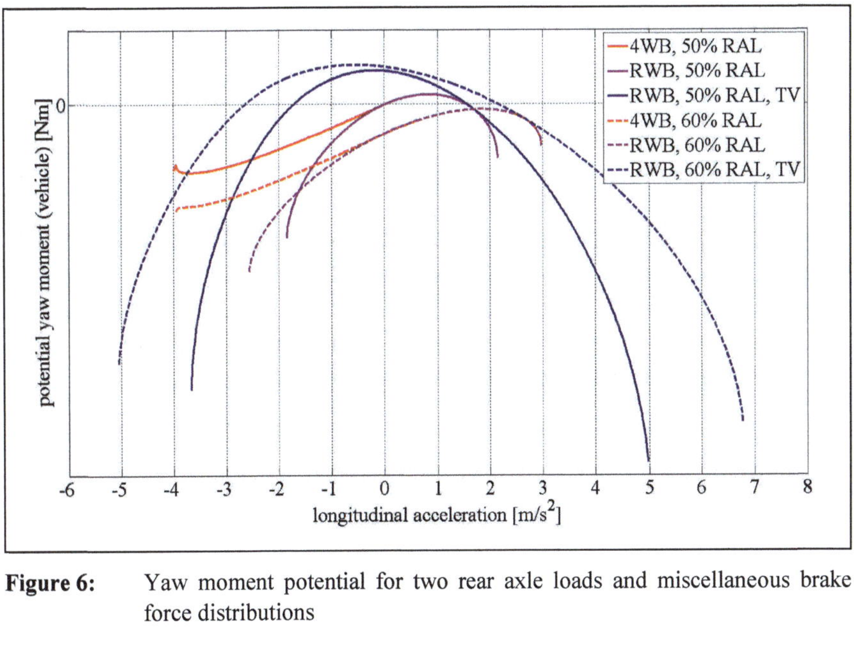

Figure 6: Yaw moment potential for two rear axle loads and miscellaneous brake force distributions

while longitudinal acceleration was varied), the resulting tire forces were calculated by paceijka's well known magic formula [11]. The friction coefficient was set to 1.0. Subsequently, the resulting yaw moment applied to the car was calculated. The yaw moment is defined to be positive when outward-turning and hence to be maximized at least for braking. Two different axle load distributions are shown: 50 % (solid lines, 50 % RAL) and 60 % (dashed lines. 60 % RAL) rear axle loads. For conventional four wheel braking with fixed brake-force distribution (4WB), a vehicle with 50 % rear axle load tends slightly towards more inturning behavior with increasing deceleration. For more rear axle load, the (4WB) car tends towards oversteering, even without longitudinal forces.

If the car brakes only at its rear wheels (RWB, magenta lines), the stability limit is already reached at lower decelerations since longitudinal forces at the rear wheels are much higher to achieve the same level of deceleration as for four wheel braking, which reduces the potential lateral forces for the rear axle. It should also be pointed out that theoretically for higher rear axle loads higher decelerations are possible. In reality, vehicle stability limits might be reached even for lower decelerations because the resulting inturning yaw moments are too large so that the vehicle will become unstable in either case and spin out.

Blue lines indicate the maximum yaw moment limits, if drive- and braking torques can be freely distributed through torque vectoring. Ones attention should be directed to the upper limits: In this driving situation it is possible to stabilize the car with 50 % rear axle load up to more than $-2\ m/s^2$ deceleration by breaking just the rear wheels. Even more remarkable are the results for higher rear axle loads: although the 'basis' without torque vectoring is more unstable than for balanced axle load distributions, the negative effects can be overcompensated.

Once again it should be pointed out that these results only show the potentials for torque vectoring in combination with high rear wheel loads. The actual realized range of stability depends on more but neglected influencing factors such as springs and dampers, anti-roll bars and kinematics. A full-vehicle analysis is shown in the next section and consolidates the results.

9.4 Simulation results for selected driving maneuvers

To substantiate previous conclusions, simulations for several driving maneuvers and rear axle loads have been accomplished. The simulation model is implemented in Matlab/Simulink and consists of a 5-body-model with 16 degrees of freedom. Tire-road-contact and resulting forces have been realized by Pacejka's well-known magic formula [11].

Three different models were investigated: uncontrolled with 50 % rear axle load, uncontrolled with 60 % rear axle load and controlled with 60 % rear axle load. Additionally, the maneuver "brake while cornering" was simulated with 60 % rear axle load and four wheel braking as well as uncontrolled rear wheel braking. The controls are realized by a feed-forward-structure which basically consists of two single track models with actual and reference behavior which then calculates the needed tire force distribution to achieve desired yawing whilst attaining the wanted longitudinal acceleration. The friction coefficient was set to 1.0.

9.4.1 Steady state cornering

To show the scope of torque vectoring systems influencing the static driving behavior, the maneuver "quasi steady state cornering" was chosen, where the car is driven on a constant circular path with the velocity slowly increasing. The results are shown in Figure 7. The original, uncontrolled car with 50 % rear axle load (RAL) serves as a reference. With increasing lateral acceleration, the steering wheel angle increases, which indicates the desired slight understeering driving behavior.

If the rear axle load is raised to 60%, the car tends more towards oversteering. Due to the fact that its axle kinematics is designed to keep the vehicle in a safe state even for high rear axle loads, the overall tendency is still understeering. Especially for lower lateral accelerations the steering wheel angle gradient is limited so that the behavior might tend to be very sensitive since little variations of the steer angle result in large changes of the resulting lateral acceleration.

To prevent this, the torque vectoring system can be used to provide understeering behavior. Notably for low and medium lateral accelerations, the driving behavior approaches the one of the car with 50 % rear axle load. For very high lateral accelerations, the understeering tendency through toque vectoring limits the achievable maximum. The second set of application parameters for heavy oversteering tendency shows even more understeering tendency than the car with

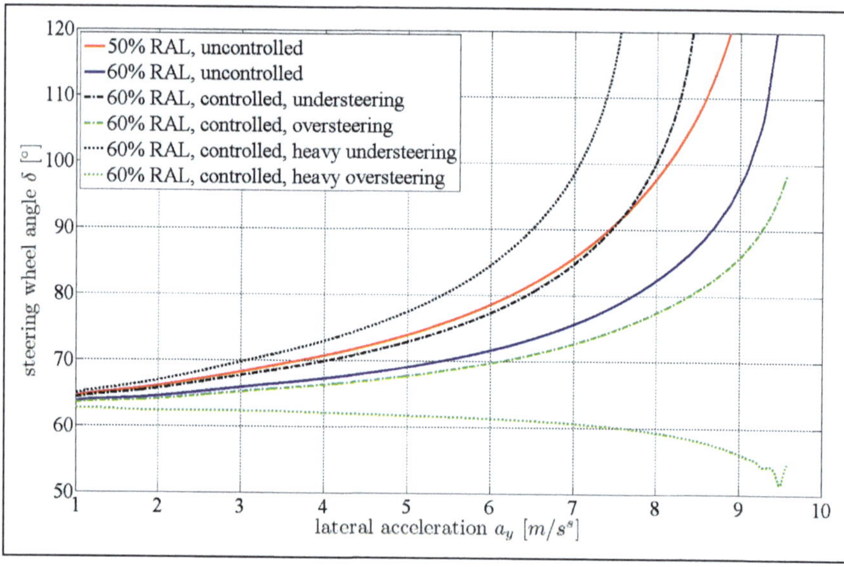

Figure 7: Steering wheel angle for steady state cornering with different torque vectoring setups

50 % RAL. It overshoots the target and hence shows solely the potential of torque vectoring.

Also, the driving behavior can be tuned into more oversteering by distributing more driving torque to the outer wheel, up to instable driving as shown in Figure 7.

It could be useful for a car which tends to understeer, but is basically not applicable for a vehicle with heavy rear axle load and rear wheel drive which itself has an oversteering tendency. Overall, one can see that the self-steering behavior can be influenced in a wide range by torque vectoring.

9.4.2 Step-Steer

Not only static, but also dynamic driving behavior can be strongly influenced by torque vectoring which is important for instantaneous response as well as for driving safety. The maneuver "step steer" is defined by driving straight with constant speed and then setting a constant steering wheel angle with high steering wheel angle gradient up to a desired height and holding it. The desired handling is spontaneous and fast increase of yaw rate and lateral acceleration

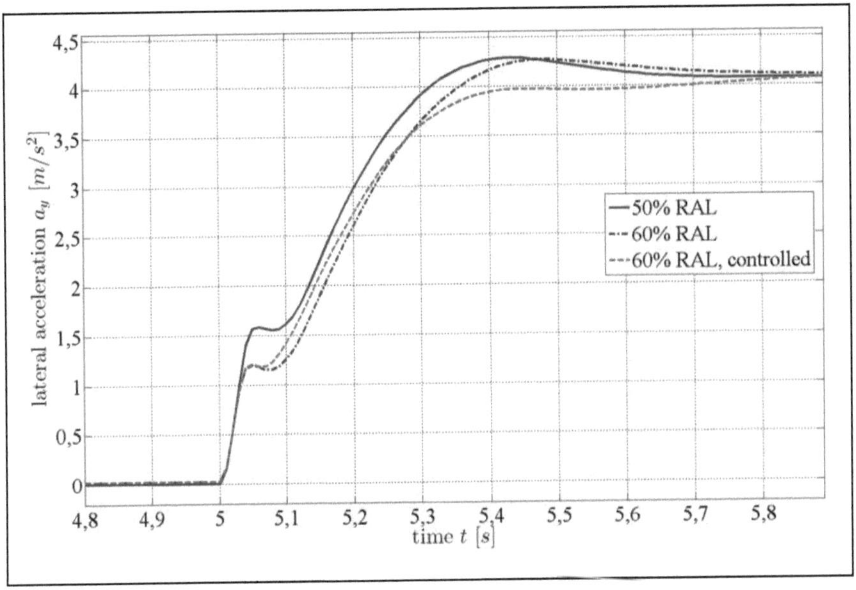

Figure 8: Step-Steer: lateral acceleration

with limited overshoot and a fast settling time. The steering wheel angle is chosen so that the resulting lateral acceleration for all configurations is the same. By the use of torque vectoring, overshoot of lateral acceleration can be eliminated as shown in Figure 8.

Figure 9 shows the resulting yaw rates. The car with 50 % rear axle load is the reference value. Higher rear axle loads without controls slow down the dynamic behavior. Also, the settling time increases. This behavior can be explained with the Hygens-Steiner theorem

$$J_z = J_{cm} + mr^2 \tag{5}$$

where the moment about the z-axis J_z is calculated with the moment of inertia of the car about an axis passing its center of mass J_{cm}, the vehicle's mass m and the perpendicular distance between the axis of rotation (which is assumed to be in the middle of the car) and the axis which would pass through the center of mass r. It is obvious that the closer the center of gravity is to the middle of the car, the smaller becomes the mass of inertia which ultimately leads to swifter system behavior.

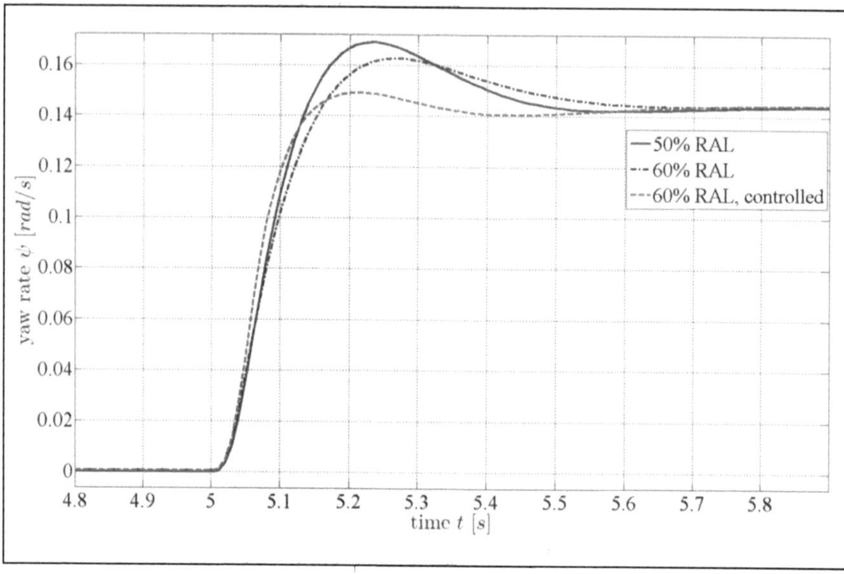

Figure 9: Step-Steer: yaw rate

The controlled vehicle shows improvements with respect to agility and safety. Although the gradient of the yaw rate is higher, overshoot and settle time are minimized.

The side slip angle is a good indicator for driving stability and safety. As it can be seen in Figure 10, the tail-heavy car needs more static side slip angle due to adverse axle loads, but the transient behavior can be optimized by torque vectoring so that the side slip angle is build up without overshoot which increases driving stability.

Figure 11 shows the control output: At the beginning of the maneuver, the car is forced into the curve. To achieve faster yawing, the inner wheel decelerates while the outer wheel accelerates, which results in an inward-turning yaw moment. To stabilize the car afterwards and eliminate the overshoot, an outward-turning yaw moment is applied to the car. The sum of driving and braking torques is zero at all times; hence there is no change of longitudinal acceleration because of torque vectoring. The car decelerates only because of cornering (no resulting overall drive torque), as it does for non-controlled cornering as well.

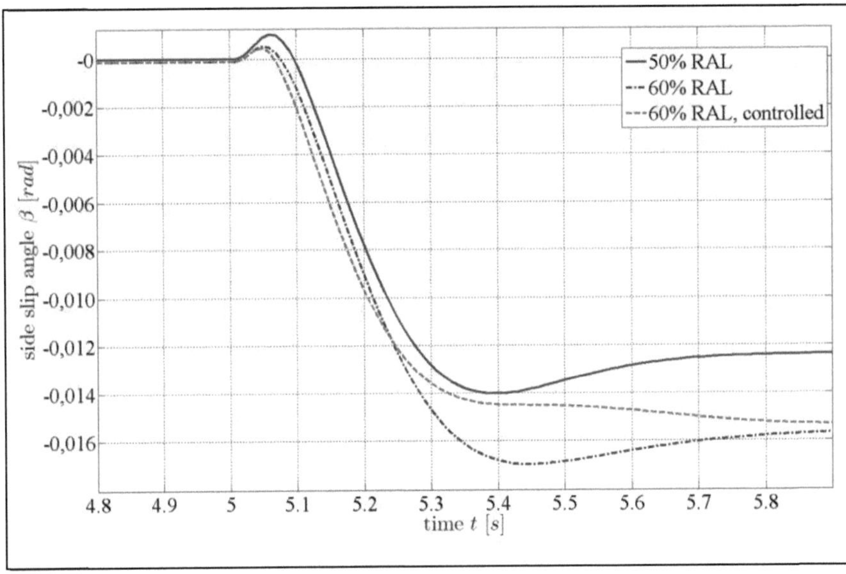

Figure 10: Step-Steer: side slip angle

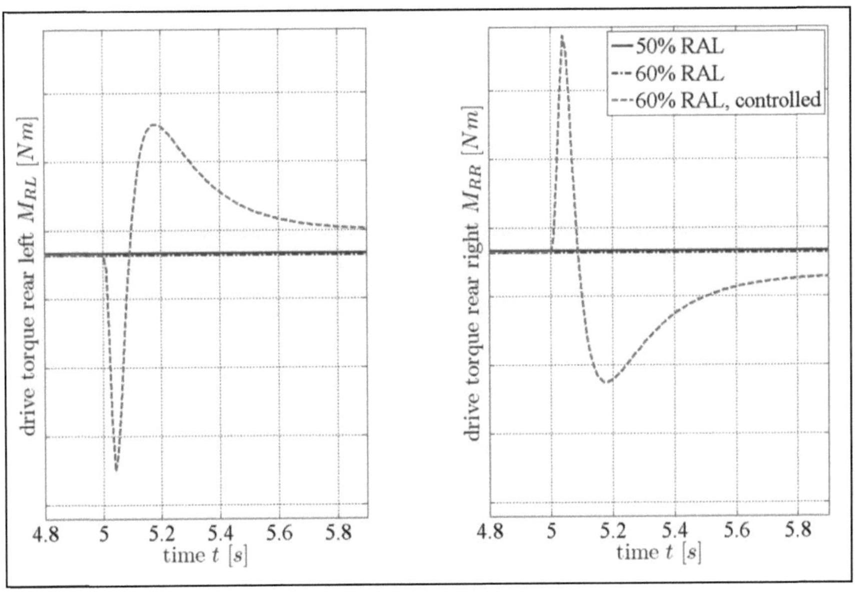

Figure 11: Step-Steer: drive torques

9.4.3 Sine with dwell

The maneuver "Sine with dwell" is an open loop maneuver, standardized by the NHTSA. It is often used to survey driving stability, for example in the U.S. to proof functionality of ESC systems. Since the drivers' input is independent of vehicle behavior, results are well comparable. The absolute value of the steering wheel angle is found via a quasi steady state cornering maneuver where the characteristic steering wheel angle $\delta_{0.3g}$ is determined. The amplitude of the steering wheel angle of the maneuver is determined afterwards via multiplication of $\delta_{0.3g}$ with a factor. This factor is set to 1.5 at the beginning and increased by 0.5 for each run. For better comparability of the different vehicle setups, the steering wheel angle is equal for the three different configurations for each run. The steering wheel angle is initiated at 80 kmph with a frequency of 0.7 Hz and a pause of 500 ms before steering back to neutral according to Figure 12. Again, the target is to achieve a fast and safe vehicle response without overshoot and a short settle time.

As for the maneuver "step steer", overshoots of lateral acceleration and yaw rate can almost be eliminated (cp. Figures 13 and 14). It should be pointed out once more that results were achieved only by feed forward controls. Most important in terms of driving stability are the side slip angle characteristics. The

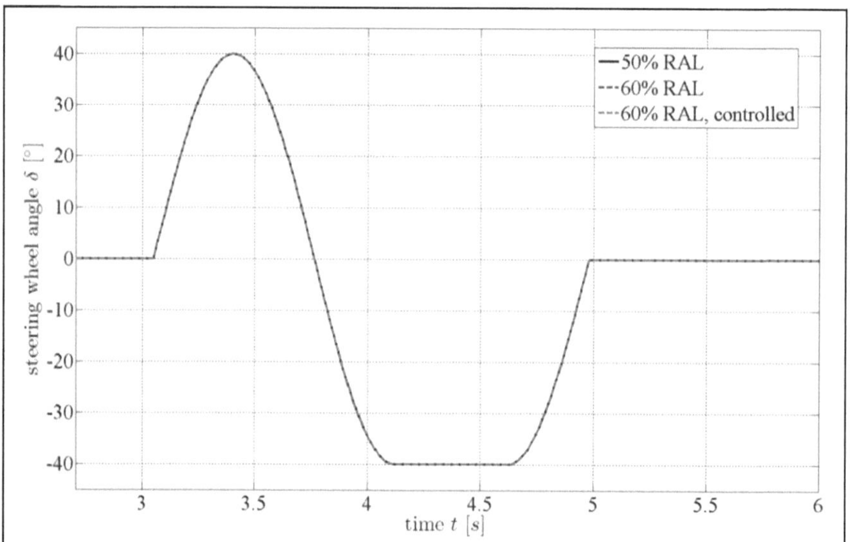

Figure 12: Sine with dwell: steering wheel angle

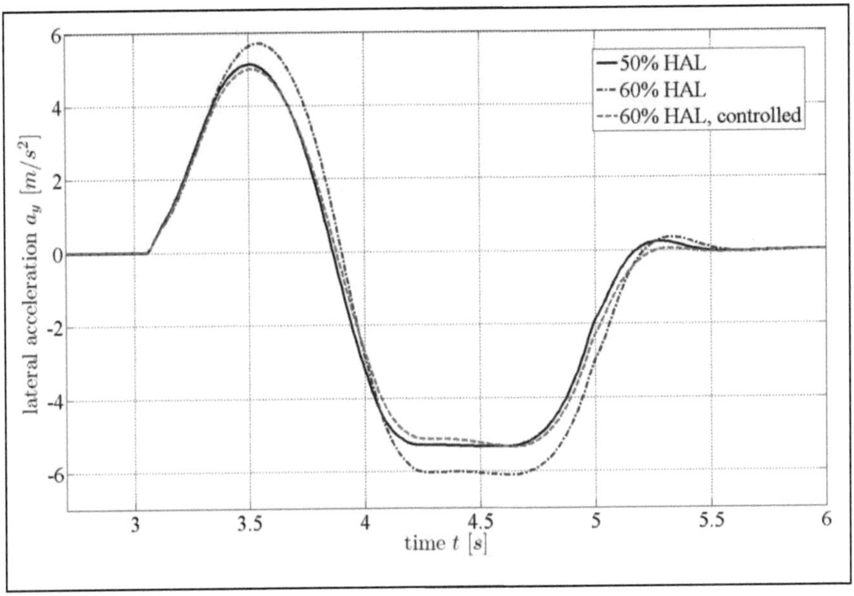

Figure 13: Sine with dwell: lateral acceleration

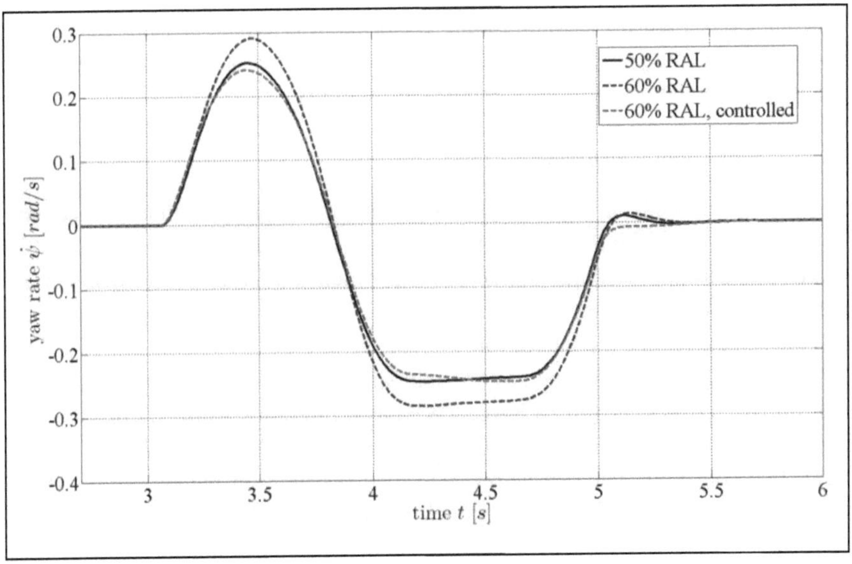

Figure 14: Sine with dwell: yaw rate

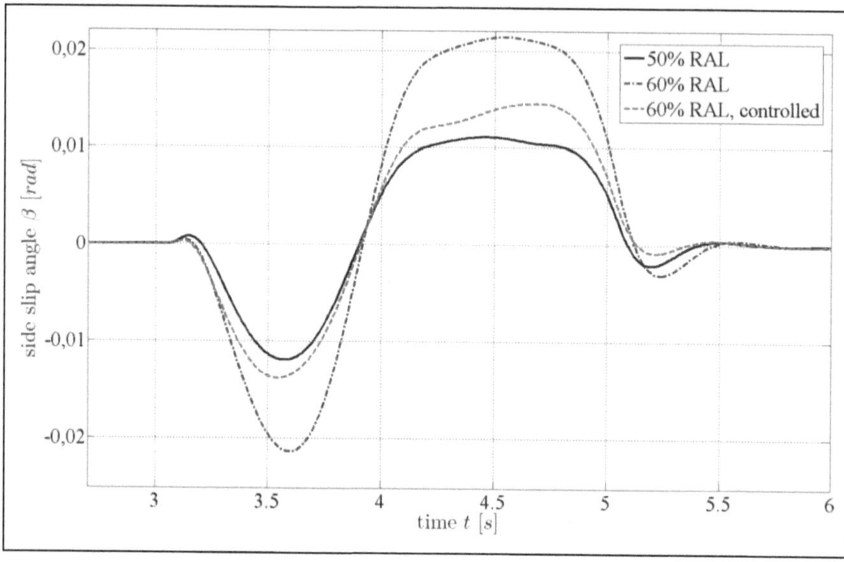

Figure 15: Sine with dwell: side slip angle

tail-heavy car without controls reaches more side slip angle than the well-balanced one (cp. Figure 15). With torque vectoring, it is almost possible to accomplish the same course of the side slip angle as for the car with perfect axle loads, which results in better drivability and driving stability.

9.4.4 Brake while cornering

An electric car needs to regenerate as much energy as possible during deceleration to achieve maximum driving range. For a rear wheel driven vehicle, this means regeneration through the rear axle, which compounds the problem of oversteering for heavy rear axle loads because the longitudinal acting forces of the rear wheels reduce the lateral force potential.

Results for four different setups are compared: A car with 50 % rear axle load and four wheel braking (50 % RAL, 4WB) as a reference value. Setup two is a car with 60 % rear axle load and four wheel braking (60 % RAL, 4WB). Furthermore, two cars with also tail-heavy axle load distribution but only rear wheel braking without and with variable wheel torque distribution (60 % RAL, RWB; 60 % RAL, RWB, controlled) are investigated. The chosen maneuver is braking while cornering with an initial lateral acceleration of 8.5 m/s^2 and desired longi-

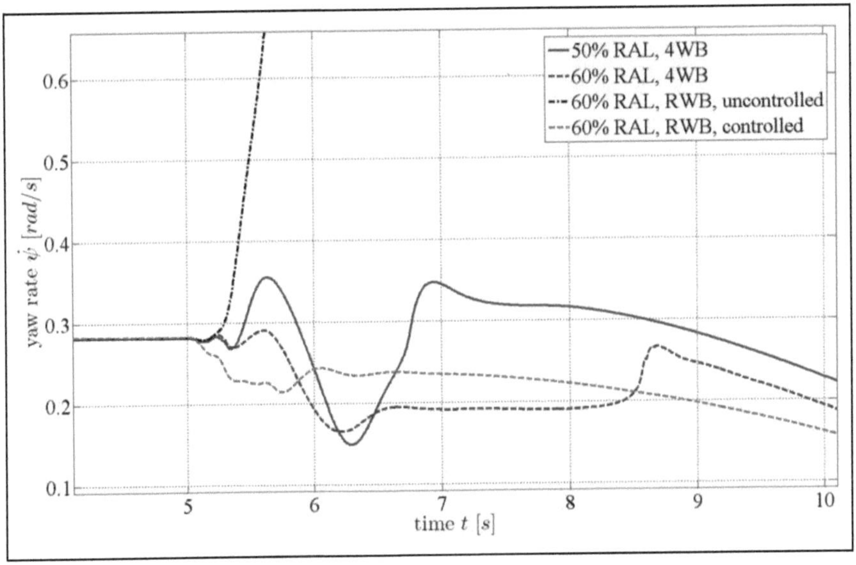

Figure 16: Brake while cornering: yaw rate

tudinal acceleration of –2 m/s^2. Initial conditions were chosen to be at the vehicle's lateral dynamics limits because in this driving situation it is most important to stabilize the car and the profitableness of torque vectoring can be pointed out best. The yaw rate indicates stability limits:

The well-balanced car as well as the tail-heavy one shows stable driving behavior with slight oscillations at the beginning of the braking if four wheel braking is applied, because the inner front wheels are closed to its maximum force potential and hence slip for a very short time (cp. Figure 16). The uncontrolled car with rear wheel braking in contrast becomes unstable and spins out, while the controlled, only rear-braked car behaves safely and even shows a smoother course than the all-wheel braked ones. Similar for the side slip angle, as shown in Figure 17: while the uncontrolled car spins out, the controlled one shows very safe behavior with even less oscillations than the all wheel braked vehicles.

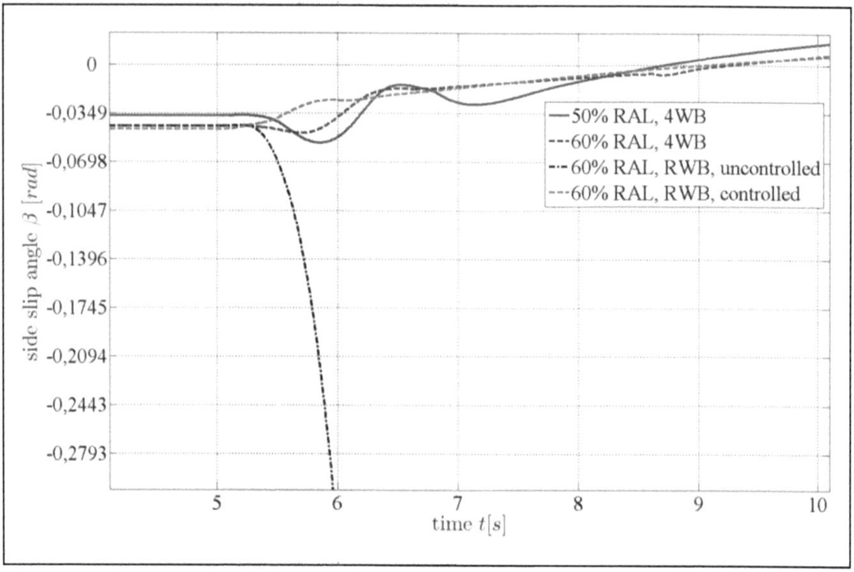

Figure 17: Brake while cornering: side slip angle

9.5 Summary

In this work, advantages through torque vectoring for a rear-driven car were shown. First, driving safety problems of tail-heavy vehicles and potentials to overcome them via free wheel torque distribution were discussed. It was shown that cars with torque vectoring at their rear axis potentially can be stabilized even at breaking while cornering. To underline the effectiveness of such a vehicle architecture, simulations of several driving maneuvers where performed. The simulations where carried out in Matlab/Simulink using a 5 body simulation vehicle with 16 degrees of freedom. As reference value, a car with well-balanced axle load was used. By raising the rear axle load to 60 %, the relevant characteristics for evaluation vehicle dynamics became worse for nearly all situations. By the use of appropriate feed-forward controls, each maneuver could be stabilized and performance could almost be adapted to the reference values of the well-balanced car. These results go along with the scientific findings of Meißner [12], who proved that torque vectoring can improve the driving behavior of a all-wheel-driven car with torque vectoring at its rear axis, as well as the proceedings of Graf, Wiesbeck and Lienkamp [10], who experienced that the recuperation potential can be increased by the use of torque vectoring.

Remarkably are the results for braking while cornering. While the uncontrolled, rear braked car spins out and becomes instable, the controlled car shows even less oscillations in side slip angle and yaw rate than the reference value.

9.6 References

[1] http://www.bmw.de/de/de/newvehicles/1series/activee/2011/showroom/gallery/indin.html#mediaID-4, accessed 11.03.2012.

[2] Mertens, A. VDE Study "Elektrofahrzeuge", contribution at „Life Needs Power", Leibnitz University Hannover, Hannover industry exhibition, 22.04.2010.

[3] Schulz, A.; Schindler, V.; Merker, S. „Elektrifizierung von Fahrzeugflotten"; ATZ, 01-2012, pp 86-92.

[4] van Meel, F; Keller, S.; Bock, T. „Elektromobilität bei Audi", 2. Automobiltechnisches Kolloquium; Garching bei München, 11.-12. April 2011.

[5] Sommer, C. „Continental study about mobility", Hannover, 15.12.2011.

[6] Braess, H.-H.; Seiffert, U. „Vieweg Handbuch Kraftfahrzeugtechnik". 5th Edition, Wiesbaden; Vieweg & Sohn, 2007.

[7] Pruckner, A. „Potentiale eines radnahen Elektroantriebs zur Gestaltung neuer Antriebs- und Fahrzeugarchitekturen"; 1.VDI-Fachkonferenz Berechnung und Erprobung bei alternativen Antrieben, Baden-Baden, 22.-23. November 2011.

[8] Haken, K.-L. „Grundlagen der Kraftfahrzeugtechnik". 2nd Edition, München; Hanser 2011.

[9] Daiß, A. „Beobachtung fahrdynamischer Zustände und Verbesserung einer ABS- und Fahrdynamikregelung". Düsseldorf; VDI, 1996.

[10] Graf, M.; Wiesbeck, F.; Lienkamp, M. "Vehicle dynamics design of the electric car Mute", chassis.tech, Munich, 07.-08. June 2011, pp 161-172.

[11] Pacejca, H. "Tire and Vehicle Dynamics"; Oxford; Butterworth-Heinemann, 2006.

[12] Meißner, T. Ch. „Verbesserung der Fahrzeugquerdynamik durch variable Antriebsmomentenverteilung". Dissertation, Technische Universität München, München, 2008.

10 The Effects of Regional Specific Requirements on the Development of Vehicle Concepts

Dr.-Ing. Alexander Prinz, Dipl.-Ing. Frank Nehuis,
Prof. Dr.-Ing. Thomas Vietor, Dr.-Ing. Carsten Stechert

Abstract: Changes in markets and technology of automobiles amplify the issue of fulfilling the requirements for as many customers as possible with a new vehicle concept. An increasing development of regional markets demands vehicle concepts or derivatives, which fit to the customer's wishes and satisfy the regional specific needs.

The approach introduced in the present publication deals with the issue of regionally specific connections between the customer's requirements and the technical concept's characteristics and properties, especially in the early stages of the product development process. Because of spreads of requirements between several markets and thus varying technical solutions, goal conflicts in the conception occur. Therefore, different strategies are shown which enable a concept solution regarding complex product structures. In the area of conflicting interests between e.g. business relevant efficiency and customer relevant diversity, a way is shown how to develop successful products on several markets.

Keywords: vehicle concepts, regional specific, sets of requirements, vehicle characteristics and properties

10.1 Introduction

In order to implement the customers′ wishes into a vehicle concept, information of regional specific differences have to be elaborated methodologically. The differences have to be considered in a number of sets of requirements. The set of requirements again have impacts on the vehicle's technical characteristics and properties. To define the best possible vehicle concept, these impacts have to be considered in the early stages of the development process. To illustrate the development of a customer-oriented vehicle, the paper starts with showing the influencing factors representing the impact from the vehicle's environment on the vehicle concept. External influences on a vehicle are, for example, the

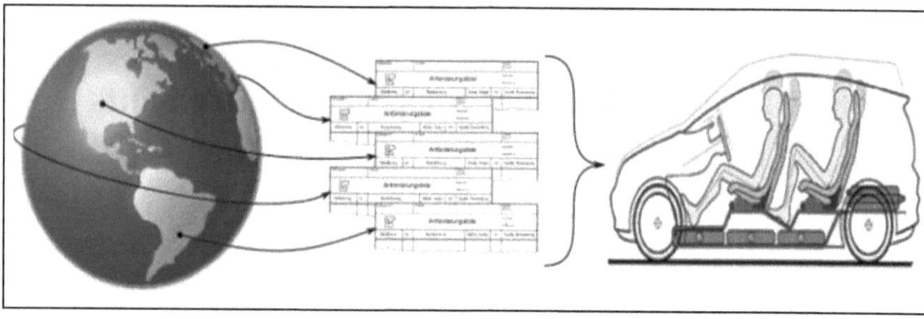

Figure 1: Illustration of the influences on a vehicle with its characteristics and properties and the resulting sets of requirements based on regional specific needs

climate in a region, the infrastructure or different cultural circumstances, [1]. As a result of only partial information on the influence of vehicle's technical requirements, some influencing factors are rarely considered in real-life vehicle development. Therefore the objective is to determine the influencing factors from the vehicle's environment and to ascertain their effect on a vehicle systematically. At this juncture, a tool can help to illustrate the influence of the vehicle's environment on his technical characters and properties. Due to the methodological elaboration of the vehicle's environment, necessary information can be collected and condensed to the regional specific sets of requirements as shown in Figure 1. Afterwards, the impacts of the influencing factors are described on the vehicle's properties and characteristics.

During this regional specific analysis, first conflicts between the requirements can already be detected (e.g. main focus region 1 "high performance", main focus region 2 "low fuel consumption"), which should be taken into account during the vehicle concept design, [2]. In contrast to this, the economical requirements towards a vehicle or model range, however, prevent from developing unlimited numbers of concepts to adapt them as best as possible to different sets of requirements. As a result, goal conflicts in the technical design of the vehicle concept occur and their solution defines the whole vehicle character. Therefore, the technical key areas of a vehicle will be analyzed methodologically in the publication. This stage reveals the amount of the fundamental technical parameters describing a vehicle and is independent from the vehicle's definite concept later on.

The mutual dependencies between the parameters are documented in a vehicle model, which illustrates the technical restrictions within a vehicle, [3]. Different concepts are caused by several values of these parameters (for example: total weight (parameter), 1.200 kg (value concept 1), 1.800 kg (value concept 2)). To establish a vehicle concept which constitutes the best possible common intersection between the customer's and the company's perspective, the overall evaluation of regional specific sets of requirements and the dependencies between the technical properties have to be considered. In order to solve the arising goal conflicts, following strategies are discussed in the publication:

1. Analyses of sensitivities of individual parameters (changing development targets/emphases in the development)
2. Optimization of an already established system
3. Use of new technologies in order to solve goal conflicts (new possibilities/ degrees of freedom)
4. Creation of several vehicle variants by strengthening single parameter values in reaction to the spreads of requirements.

Describing the resulting set of solutions aims at leading to technical and ecological promising concepts with high market potential, [4].

10.2 State of Research

10.2.1 Integration of the vehicle conception into the process of technical design

In order to develop a complex product, the definition of requirements for the system or for the system's single objects is a central stage to fulfill the customers' needs. This becomes clear by considering established design methodological procedures, in which "clarifying and specifying the task" represents an independent procedural stage and features the compilation of technical requirements, [5].

In the process model according to the VDI Guideline 2221 the task is clarified at the beginning of the development process. The needs and wishes required from a product are defined in consideration of the boundary conditions. The result of this stage is a list of requirements in which all requirements are collected, exemplarily illustrated in Figure 2.

The specifically determined values of requirements and data regarded to a customer or a group of customers define the so-called set of requirements and represent consequently a first model of the product. In the development process,

Client: Institut for Engineering Design		Product: Vehicle concept			Date: 11/28/2011 Sheet No: 1
INSTITUT FÜR KONSTRUKTIONSTECHNIK		List of requirements			Processor: Nehuis Revision No: IK-001
Structure	**No.**	**Description**	**Values, data**	**Type**	**Source, remark**
1. Development	1.1	Avoid overlaps with patents	all patents	F	Dep. of development, Mr Sketch
2. Production	2.1	Maximize number of items	n > 400.000/a	M	Dep. of sale, Mrs Value
3. Usage	3.1	Reachable maximum speed	v_{max} > 220 km/h	M	Dep. of marketing, Mr Spot
	3.2	Increase the life time of all components	s_{min} > 50.000 km	M	Dep. of development, Mr Sketch
4. Recycling	4.1	Enable to devide material of assemblies	homogeneous materials	W	Dep. of environment, Mrs Flower

Figure 2: Section of a requirement list with one set of requirements for a vehicle concept (example)

the product which has to be developed will be concretized stage by stage. As a result, further information is gained and also included into the set of requirements, [6]. However, already defined requirements may have to be changed during the development process. This is the case if the fulfilling of a requirement is only possible e.g. with increasing knowledge of production or if some boundary conditions and restrictions are unknown at the beginning of a development process [7]. If the requirement list has to be modified in the later stages of the development process, it is generally associated with high costs and increasing time pressure. Especially requirements which were already determined in the concept phase of the product can only be changed with raised effort, Figure 3.

In order to reduce the development costs, product changes in the later stages of product life cycle have to be prevented. Therefore, new methods are used, which raise the level of knowledge in the early stages of the product development, [8]. This means, regarding to “clarifying and specifying the task”, that the requirements on a product have to be considered carefully, and for this purpose first information of the product and its environment should be known. It thus is advisable to spend enough time in determining the requirements in an early stage of the developing process whereby initially more time is needed. Overall, developing time and costs can be reduced generally by an intensive task clarification.

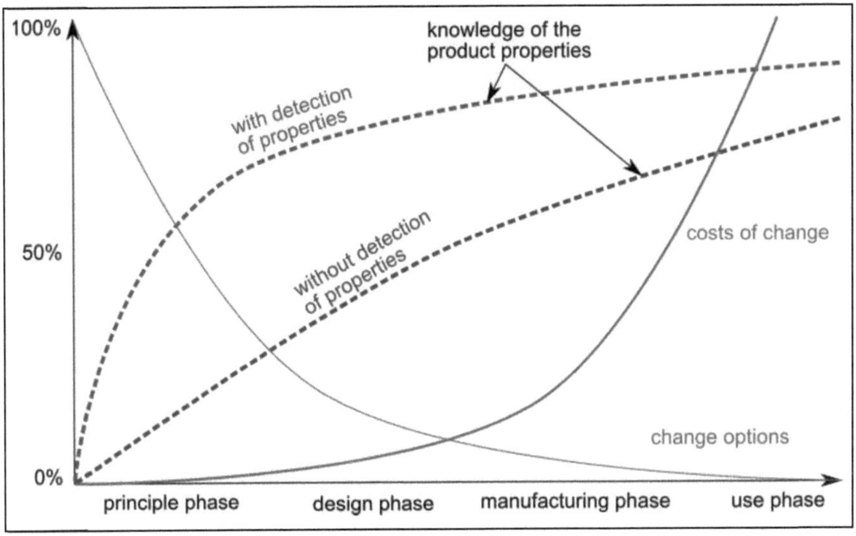

Figure 3: Reducing modification input and costs by early recognition of the system properties, regarding [8]

10.2.2 Embedding the system "vehicle" into the system context

In order to set the requirements on a vehicle as comprehensively as possible at an early stage of the developing process, the vehicle's environment has to be extensively analyzed. A complete overview over all influencing factors is hard to be accomplished in real life because a large number of influencing factors have an effect on a vehicle indirectly as well as directly. The created complex system is illustrated schematically in Figure 4, [2].

Besides the knowledge of the influencing factors which have an effect on a vehicle concept, the engineer has to have fundamental system knowledge. This helps him to recognize the effects of requirements on the vehicle concept and its components and parts. Further, it allows him to determine the requirements on a concept consciously. If there is no fundamental system knowledge, the engineer may determine requirements which are not or only partly realizable. The vehicle's environment, shown schematically in Figure 4, directly effects on the vehicle concept's requirements.

Environmental facts can be boundary conditions of a company or a country´s economy, specific wishes of a customer or regional specific varying climate

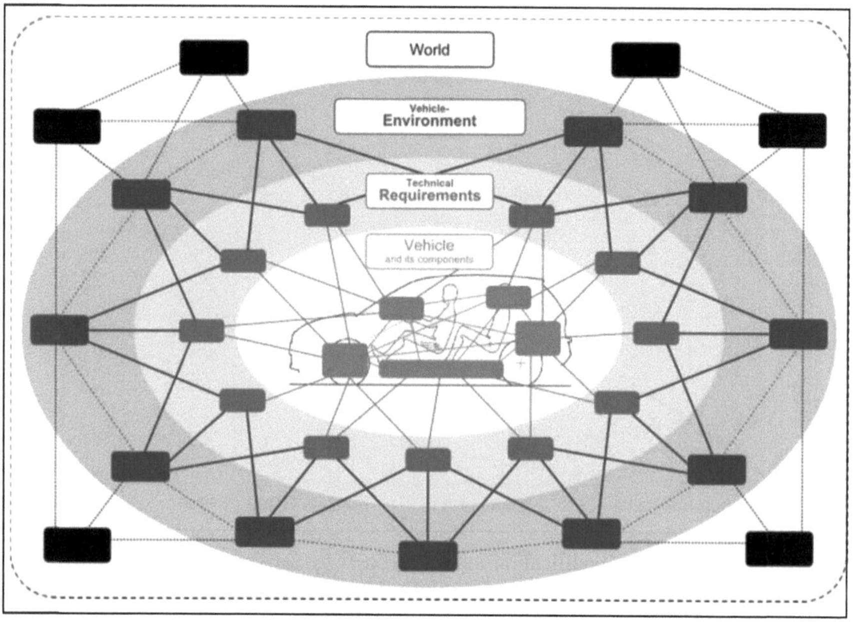

Figure 4: Schematic illustration of the vehicle and its embedding in the overall vehicle environment

conditions which define the vehicle concept in the first stage. Thus, for example, the humidity in a region has an effect on the technical requirements of a vehicle by ensuring a corresponding corrosion protection on the product to be realized. Requirements the vehicle is provided to fulfill, again have an impact on the vehicle's parameters (characteristics and properties). The relation between the technical characteristics and properties, the requirements towards the vehicle and the surrounding vehicle environment is illustrated schematically in Figure 5.

Usually, a requirement does not only have an influence on one single vehicle's characteristic and property, but on several ones. Thus, goal conflicts may occur when realizing the requested set of requirements.

Furthermore, various requirements may have influence on a characteristic or property. Technical requirements like the warranty of corrosion protection and simultaneously the minimizing of total weight can be claimed by the customer and/or the company. If the engineer raises the protective varnish in the painting, there is a goal conflict regarding to minimize the total weight, [9].

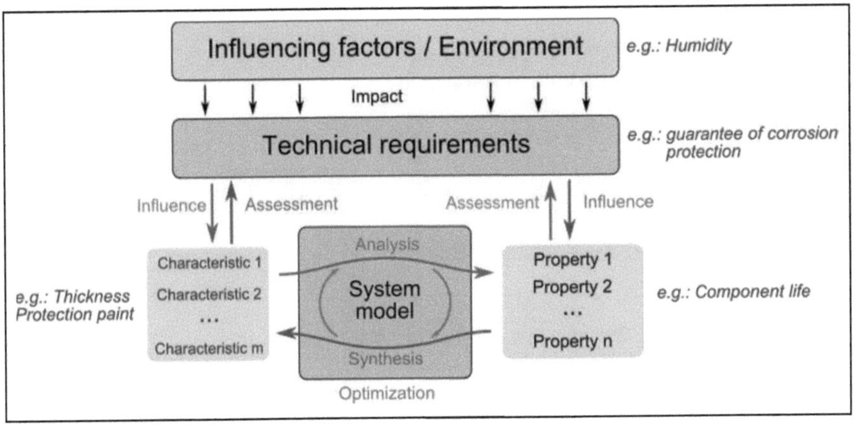

Figure 5: Illustration of the relation between influencing environmental factors, technical requirements and the characteristics and properties of a vehicle concept

10.2.3 Technical requirements of a vehicle concept and placement of the system "vehicle" in its environment

Considering the vehicle's environment in terms of the product life cycle, a large number of influencing factors can be defined. These include obvious influences like infrastructure or legislation, but also factors such as the vehicle owner's individuality or culture in a region, [1].

Already in the 1970s, the task of investigating the product environment has been recognized in the methodology of technical design. One approach was to provide methodological helping tools for systematical requirements compilation, e.g. as general checklists [10] or morphological search procedures [11]. In order to facilitate a target-oriented, product-related clarification, later on these methods were continued computer aided, with a branch-specific level of detail, [12]. These suggestions didn't find a wide range of practical use yet, but specially adapted software tools can probably change this. They have to be more user-friendly compared to the existing tools, maintainable by the product specialists and need to be loaded with specific product knowledge in a significant extend, [13]. In order to bridge the scientific gap in this publication, the considered influencing factors were compared to the technical requirements in a relation system and the respective impact constituted by connections.

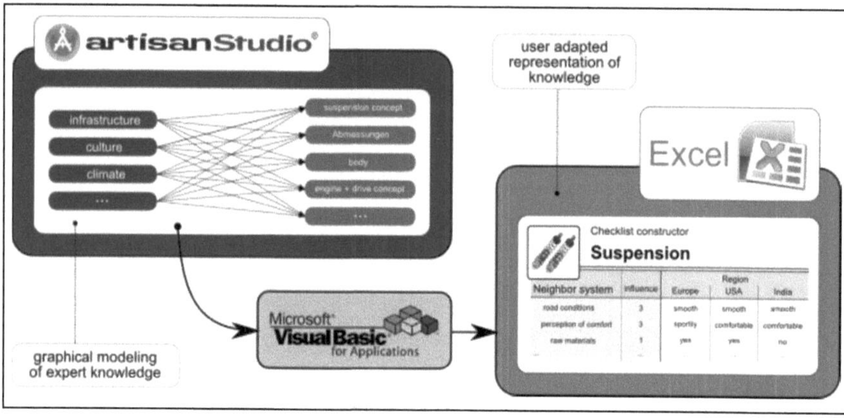

Figure 6: Scheme of knowledge compilation adapted to user expertise

To handle the relations between the characteristics, the properties and the influencing factors, these elements are documented in the System Modeling Language (SysML). SysML is based on the Unified Modeling Language (UML), a programming language for modeling complex systems. In order to simplify the handling with SysML, the SysML-Editor "Artisan Studio" is used additionally. This enables easy graphical set up and the extension of using comprehensive systems within the SysML, [14].

The empirically ascertained data is time-dependent and reflects the current status in a region, in opposite to the technical requirements, the influencing factors and the relations, [15]. Because of the systematical analysis of the vehicle's environment as well as the software realization in a relation system, the basics for the conscious definition of technical requirements on a vehicle concept have been done. Subsequently, the effects of the influencing factors on the vehicle's properties and characteristics are demonstrated, whereby regional specific sets of requirements are becoming obvious. In that way, all influencing factors and its effects, as well as the information for several regions, can be constituted for particular technical requirements, illustrated in Figure 6.

The user-centered output of information makes it easier for the development engineer to define the technical requirements for the vehicle concept. Thanks to the weighting of relations, the engineer can also estimate which compromise he wants to make when defining the requirements based on influencing factors. The documentation of various regions in the relation system allows the identification of technical requirements, in which regional specific differences – so-called spreads of requirements – occur.

10.2.4 Interaction between the characteristics and the properties of a vehicle concept

During the work on a vehicle concept, the "level of elaboration" and "level of detail" have an importance as independent coordinates of the technical design space. The level of elaboration describes the amount of data known about the concept. The level of detail means the type and number of relevant interactions and links between the technical characteristics and properties of the concept.

Because of the special challenge, there is an interest in the case of application in which no predecessor model for the vehicle to be developed exists, and therefore no data is known at the beginning. This case is characterized by a special connection between the level of elaboration and the level of detail. At the beginning of the technical design process, the grades of detail and elaboration are at a low level, the concept is a so-called rough attempt. Usually, more exact statements are done stage by stage, so the number of considered vehicle properties increases.

This causes the level of detail to soar and to rapidly reach a high level. If the level of elaboration is finally well advanced, the level of detail grows only slightly. In this stage, the major part of links already got involved and there is only a little number of new relationships coming into account. In such a "late stage", changes of characteristics should be small because every variation causes a huge number of impacts. Figure 7 illustrates the hypothesis for such a process.

From the little knowledge regarding the vehicle concept a "conceptual uncertainty" arises and is unavoidable for the concept engineer at the beginning. That is why a low level of elaboration helps to get ahead, because it implicates an assessable level of detail. This is the guaranty for the first steps in the vehicle project and the basis for the stagewise refinement later on. Because of that, the support in the first stage (section I in Figure 7) has certain importance; appropriated tools have to be provided.

A key element is the model of a vehicle, which contains the essential technical characteristics and properties and also their links. In the notified early stage of the concept development, there are different opinions in terms of the useful amount of vehicle parameters. To make clear which state is used in this publication, Figure 8 shows the parameters in the concept. They are put together in

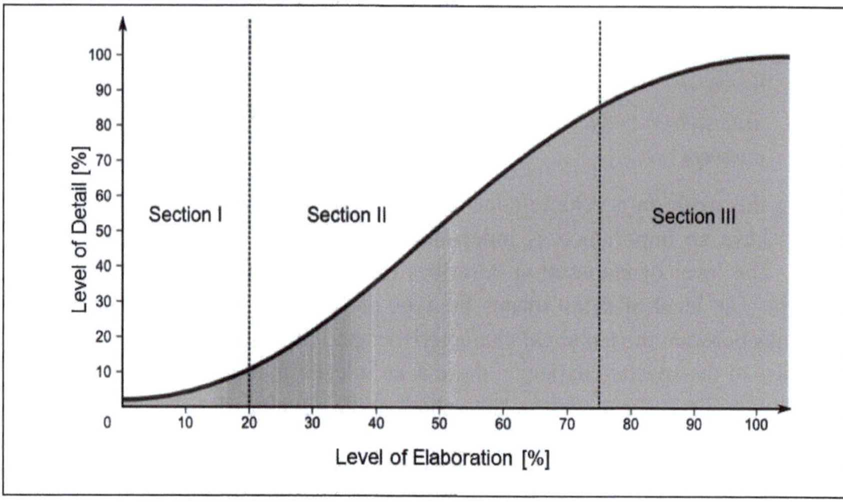

Figure 7: Connection between level of elaboration and level of detail in the development of a vehicle concept (hypothesis), referred [3]

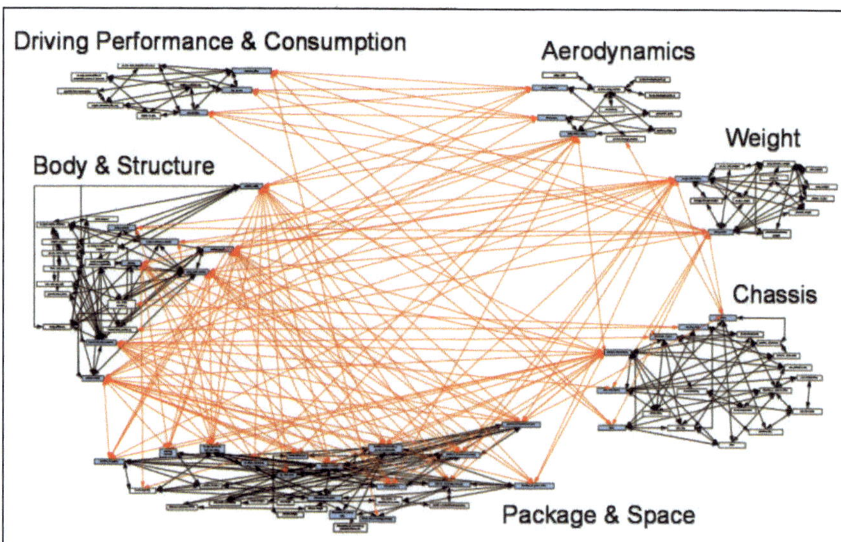

Figure 8: Fundamental vehicle parameters including the main interactions, [3]

the main groups "driving performance & consumption", "aerodynamics", "weight", "chassis", "package & space" and "body & structure". Concept characteristics like the layout of the axles or the entry concept are included as well as technical properties (e.g. drag coefficient, total weight and vehicle length). Whether as characteristics or properties, parameters have to describe a part of the vehicle. They have to be changeable in their value, at least in a way using other direct influences. All characteristics and properties result in a sufficiently detailed description for the first approach of a vehicle concept.

In addition, according to real-life technical relationships, no characteristic stands entirely isolated for itself. Dependencies occur both within the main groups and between parameters of different main groups. If the value of a single or of several parameters is changed, the concept changes as a whole. This happens on the one hand through the parameter value itself, on the other hand by influencing and thus changing of linked parameters and their values.

10.3 Goal conflicts in the vehicle design

During the development of the complex vehicle system, a huge number of different goal conflicts occur. A goal conflict, referred to [16], occurs, if various influences have an effect on a single characteristic's value and they want the value to be changed in the opposite direction. A conflict between two requirements exists if the fulfilling of the first requirement leads to a reduction in fulfilling of the second requirement. In real life, it is difficult to have an overview over the entire network of characteristics and properties. Because of that, goal conflicts have to be identified computer aided. Due to the computer aided tools the characteristics and properties can be influenced in an optimal way to fulfill the vehicle requirements.

In this chapter, possible connections in the whole vehicle system are shown and typical goal conflicts in the system are presented. The relationships between the technical characteristic "vehicle length" and the vehicle properties "drag coefficient" as well as "total weight" are used as an example.

10.3.1 Relationships between vehicle and vehicle environment

In the development of vehicle models, regional specific requirements like different customer wishes, legal provisions or variant environmental conditions compared to the European market, were conditionally considered yet, [1]. Because of

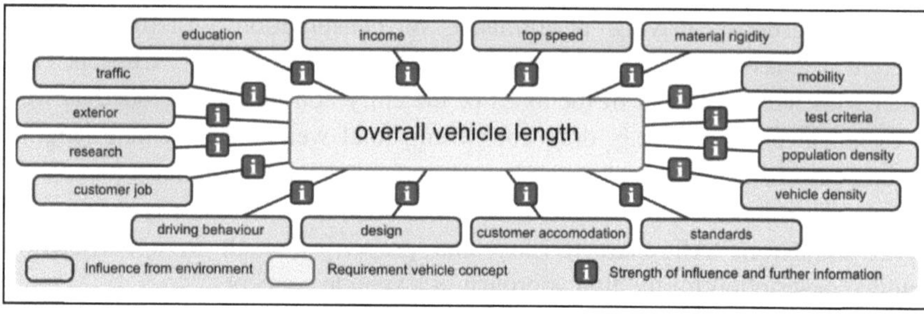

Figure 9: Representation of the influencing factors out of the environment which affect the overall vehicle length

the increasing customer needs for individual mobility, e.g. in the BRIC-countries, it is becoming more and more important to define the requirements on a vehicle in various regions and to establish the resulting sets of requirements. In this case, a clear and comprehensible example is the climate in a region. The average daytime temperatures in Rockhampton, Queensland in Australia, from January to December are between 16,2°C and 26,7°C, whereas the temperatures in Werchojansk, Russia, vary between –47,0°C and 15,2°C in the same period of time (source: German Weather Service). Besides this example of regionally different climate conditions, also less trivial influencing factors, e.g. the habits and customs in a region, may have an influence on a vehicle. This will result in various values of technical requirements (e.g. focus region 1 "high longitudinal acceleration", focus region 2 "low consumption"), which should be considered in a vehicle concept, [17].

The computer aided vehicle environmental model described in Section 10.2.3 makes it possible to show the influencing factors affecting a vehicle concept with an acceptable effort. In addition to the pure presentation of the influencing factors, it is possible to attach additional information to each influencing factor. If the designing engineer defines the characteristic "vehicle length", the effect of various influencing factors of the vehicle's environment on the technical characteristics and properties will be constituted by the help of the tool. This clarifies the regional specific situation. This specific connection of knowledge is shown schematically in Figure 9.

The vehicle length is influenced by several factors. One influence resulting from the geometric relation to the vehicle´s footprint comes from the population density and thus from the density of traffic in a region. In large cities of Germany with population densities of about 4.300 citizens per km² in Munich and about 3.900 citizens per km² in Berlin (status 2007) the vehicle footprint plays a comparatively minor role. On the other hand, the importance of the traffic space used by vehicles in urbanized regions obviously increases. In that case, for example, the megacity Peking has a population density of about 22.400 citizens per km² (status 2007), which has to be considered if a car should be developed for a large group of customers. In this context the so-called Kai Cars, which are especially developed for the Japanese market, should be mentioned, which have a particularly small vehicle footprint and therefore are considering the circumstances of urbanized regions.

The material strength as another example does not obviously have an impact as an external influence on the vehicle length. But the body stiffness plays an important role for driving and is, among other ways, expressed by the lightweight index. Furthermore, a higher torsional and flexural stiffness of the body leads to a better crashworthiness. In order to get a similar flexural stiffness in a long vehicle compared to a short one without any shape optimization, the stiffness of the used material has to be increased. That might bring higher costs resulting from a higher material quality, which is in conflict to the claimed low total costs of the vehicle.

10.3.2 Goal conflicts in the system itself

Besides the goal conflicts between the environmental influences and the vehicle concept, also unwanted changes of characteristics and properties are to be expected during the definition of the respective parameters. This is because of the links and therefore the coupling between the characteristics and properties, as already shown in section 10.2.1. They lead to a reduction of degrees of freedom in designing the concept. On the one hand this effect is on purpose, in order to increase the level of elaboration (see Figure 7). On the other hand, it may lead to a technical solution far beyond the desired requirements or even to impossibility. To this, Figure 10 illustrates the tension between the drag coefficient and the total weight, if the length of the vehicle changes.

The vehicle length interacts directly with the body shape, which should have a drag coefficient and a facial area as small as possible. The ideal drop shape features a special relation between width and length. If a certain width cannot be underbid because of the passengers or other influences, it requires a bigger length for the ideal flow behavior. The vehicle weight runs in contrast to that and also has a major influence on driving resistances and therefore the energy needs.

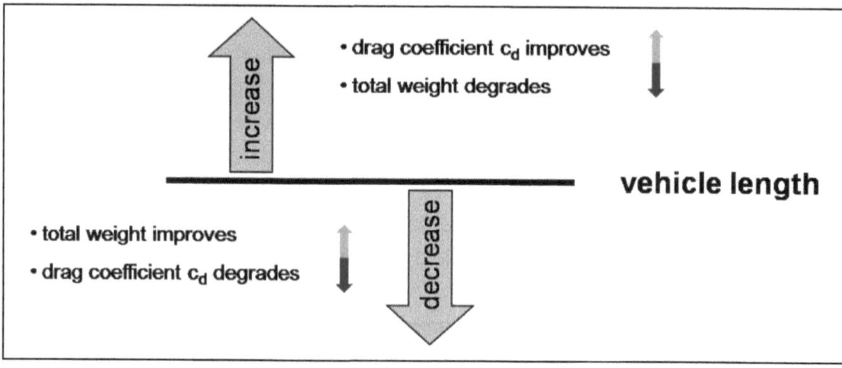

Figure 10: Goal conflict between drag coefficient and total weight as result of changing the vehicle length

Among other measures, a reduction of the vehicle's dimensions contributes to a reduction of weight. This is based on the used material itself, but also on stiffness. If a vehicle becomes larger the supporting structures have to absorb more forces, which are related to additional vehicle weight by enlarging the cross sections of beams or other measures.

In literature, various tactics to solve technical goal conflicts are mentioned. Besides the insert of an additional degree of freedom, basically two fundamental principles can be found: the compromise and the prioritization of one side. The second principle requires a decision in favor of a single optimization direction, whereas the negative development of the other direction is considered less important. However, the compromise requires the finding of a local optimum, in which both criteria reach acceptable values. This kind of solution is also known as Pareto-Optimum, referred to [18]. For the special case mentioned in Figure 10, Figure 11 illustrates how a compromise solution may look like. The diagram shows the graphs of the total weight, the drag coefficient and the fuel consumption, each as a function of the vehicle length.

On the basis of such a graphical solution of the goal conflict, the solution set is easier to recognize. But depending on the set of requirements and the boundary conditions, the solution space can be limited. Such a relatively simple and "manual" compiled solution requires several fundamentals. On the one hand, the description of the system "vehicle" has to be on a manageable level as assumed in this case. In addition, a set of requirements is necessary, which can be linked to the technical characteristics and properties and presents a clear prioritization of requirements.

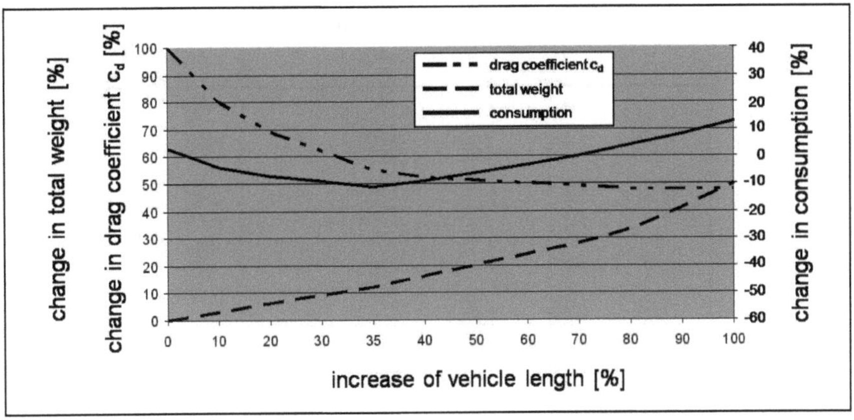

Figure 11: Solution of the goal conflict between drag coefficient and total weight

10.3.3 Resulting sets of requirements

Global operating companies in the automotive industry have to satisfy the wishes of both established and developing markets. In that case, a singular direction of product development (a single vehicle for all markets) seems to be increasingly inappropriate to fulfill the wishes and needs of customers sufficiently. This statement is underpinned by an overview of the best-selling vehicles from an exemplary month of the year 2011 for the markets Germany, India and the USA (Figure 12).

Germany		India		USA	
vehicle	length [in mm]	vehicle	length [in mm]	vehicle	length [in mm]
1 VW Golf	4199	1 Maruti Alto	3655	1 Toyota Camry	4805
2 VW Passat	4771	2 Mahindra Bolero	4107	2 Toyota Corolla	4371
3 VW Polo	3970	3 VW Vento	4384	3 Ford Fusion	4013

Figure 12: Best-selling cars in Germany, India and the USA (status 06/2011)

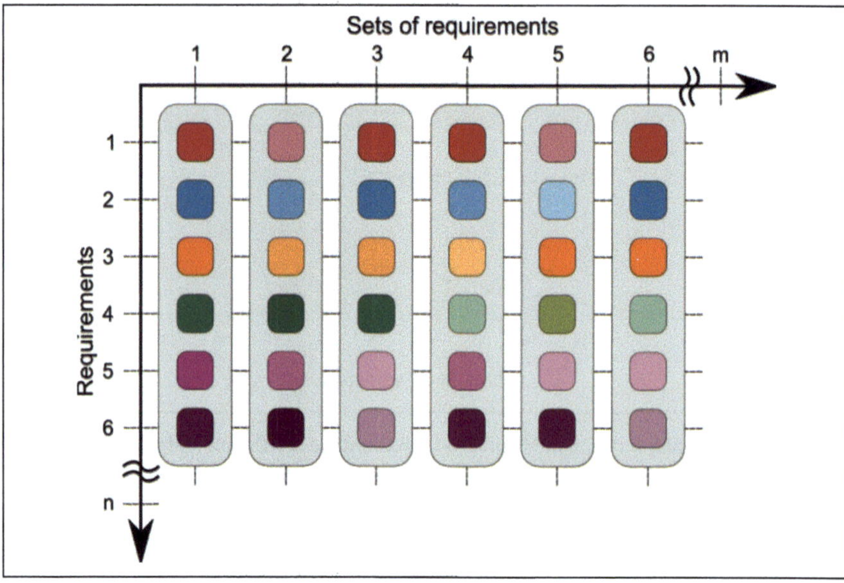

Figure 13: Schematic illustration of various requirements in different sets of requirements

It can be noted that the main parameters of a vehicle (e.g. number of seats, luggage space, price, maintenance costs) are certainly similar, but the parameter values differ significantly (e.g. trunk volume = 250 liters vs. trunk volume = 550 liters) between the markets and between various groups of customers. In that case, it is a so-called spread of requirements. From the spreads of requirements result particular collectives of specific requirements and their values, which are documented in the sets of requirements. Each set includes the values of requirements for a customer region like in an ideal case. Figure 13 clarifies this approach schematically by illustrating the various values in the sets of requirements by various color intensities in a row.

Taking a closer look on the sets of requirements in theory, the customer group represented by a set of requirement should be small. In that case, requirements can be adapted to the customer's needs in a better way. The extreme example is an individual set for every single customer worldwide (extreme I). The industrial logic of huge companies pursues the contrary direction, mainly because of costs and reasons of complexity. For this reason the companies try to identify the common interests of large customer groups in order to fulfill the needs of many customers with a single product and to produce large quantities of it. The extreme in this consideration is a single set of requirement for all customers and markets on the world (extreme II). While the effort to spend for evaluation, implementation and keeping up-to-date of "extreme I" prevented the realization, at least some approaches were made to develop a "world car" for the purposes of "extreme II" in the past. These approaches, however, did not work as successful as intended, so they were not confirmed.

Therefore, it can be concluded that a useful solution in between the extremes has to be found. In that case, the company's boundary conditions (development capacity, production sites, product portfolio etc.) have to be considered and at the same time a major part of customers′ wishes have to be represented in the sets of requirements. Here the Pareto-Optimum (see Section 10.3.2) should also be referred to because it is expedient in terms of an overall economic perspective to fulfill the wishes of a group of customers as huge as possible with a finite number of products, which can be produced economically. Another dimension, which has not been taken into account yet, is the time. The requirements out of the vehicle's environment (regions, customer groups…) also depend on temporal changes as well as the companies' possibilities or other boundary conditions (laws, guidelines, scientific evidences).

In the last years, an important change on the world markets could be noticed (keyword: "Emerging Markets"). These changes generate new customer groups and new sets of requirements. Similarly, a lot of companies created additional degrees of freedom by establishing strategies of using unified modules for their new cars. That aims at providing economic vehicles even in small quantities, individually tailored to the customers' needs. Finally, the customer wishes also changed, triggered by social changes, through scientific evidences (keyword: CO_2 discussion) and other influences. Therefore, sets of requirements are always mapping the current status. Similar to the number of sets, also a bandwidth in time exists, from updating a particular requirement in "real-time" to requirements with multiannual revision periods. In both cases the data has to be determined regularly and has to offer a possibility for a look into the future. From the strict projection of trend lines through to scenario techniques including the calculation of trend-breaking events (so-called Wild Cards), a large number of possibilities exists, which will not be examined in more detail in this paper.

10.4 Possibilities for analyzing and solving goal conflicts during the concept design of a vehicle

In the previous section, various goal conflicts were discussed. Goal conflicts are affected by various environmental influences as well as by technical properties during the definition of the requirements. All the goal conflicts have an effect on the determination of the vehicle's characteristics, which are set by the design engineer. The engineer has to find an adequate compromise between two conflicting characteristics in order to design the vehicle concept as best as possible. Therefore, he has methods at his disposal, e.g. the analysis of sensitivities or tools for optimizing conflicting characteristics.

If, however, a spread of requirements of too big to handle is detected in the set of requirements or if two conflicting vehicle characteristics cannot be considered in a proper way, the goal conflict can only be solved by a fundamental changing of the system. Through the employment of new technologies the design engineer may get new possibilities to influence the existing goal conflicts in a positive way. But if there aren't new technologies at his disposal, an adequate segmentation of the vehicle concept has to be considered. In this case, also the economic efficiency of different vehicle variants or the boundary conditions within the company have to be consulted. The following parts of the paper describe the mentioned possibilities of influencing a vehicle concept by handling the goal conflicts.

10.4.1 Examination of single characteristics` sensitivities

The relations between the influencing characteristics and properties of a vehicle can be described in a complex relation system. In this relation system the relations can generally be characterized quantitatively by a formula as well as qualitatively by coherences, for example by using a knowledge base. The knowledge about the interactions between the characteristics and properties, which is stored in the relation system, is the basis used by the design engineer for changing a vehicle characteristic in a sensible way. In order to gain information about single values of the set of requirements and its influences on the overall system, their sensitivities on other system characteristics have to be considered.

Therefore, in the context of sensitivity analyses, at least two values are put into relation. These analyses are procedures showing the change in an output variable y, caused by variable x influencing variable y. In contrary to the determination of key numbers, during a sensitivity analyses the other system characteristics are allowed to change with the considered way the system would do as a whole. Thereby, the sensitivity of the considered characteristics is in focus, [19].

As an example for the sensitivity analyses, the impact of the possible range of a purely electric driven vehicle concept on the total vehicle weight can be shown. It is obvious that the total weight increases exponentially with the vehicle's range. It can be explained by the fact that with additional range more battery capacity has to be provided. The outcome of this is an increasing battery weight. Because of the higher battery weight, however, other vehicle components like the suspension have to be adapted. Hereby a weight helix is set into operation, which reflects in an exponential progress.

10.4.2 Optimizing the vehicle's characteristics according to the target criteria from the set of requirements

Besides the sensitivity analyses of single characteristics and properties, also optimization methods are provided to define the vehicle's parameters. These methods enable the engineer to define the vehicle characteristics arithmetically.

The application of optimizing methods is particularly necessary during the concept phase, when a manual definition is impossible because of the system's complexity. A various number of different combinations can be determined between characteristics, properties and goal conflicts working in opposite directions. Further difficulties come from the non-linear dependencies between characteristics and properties, whereby the influences of a parameter on the vehicle are hard to imagine. In order to set the vehicle's characteristics in an optimal way, the relations between the system's characteristics and properties have to be known. If these relations exist in a model, it is possible to get a quick overview of the expected parameter values regarding important requirements with the help of a computer, [20].

Therefore, the relations in the vehicle have to be transformed into a mathematical model describing the relations between the characteristics and properties with sufficient precision. Because of the mathematical transformation, the optimization can be computer aided. In fact, finding an adequate mathematical description usually is a difficult and complex task. If a mathematical description cannot be found, the relation between the characteristics and properties can also be illustrated by a computer aided simulation model. In that case, every system can be analyzed and optimized through computer aid.

Important criteria to transform the system into a model are the stage in the development process as well as the operating time for optimization. At the start of the development process, an optimization model should point out tendential results in a preferably short processing time. The need for more exact results increases in the ongoing of product development continually because of the requirements regarding the liability of the statement. This fact generally has a negative impact on the calculation time for optimization. An optimization model features characteristics, boundary conditions and goal conflicts. Depending on the process of optimization, the model is evaluated iteratively. In each evaluation, new optimization parameters are detected. The new parameters are used in the optimization model, and finally the new model is assessed again. In the further course, the optimization process is searching for other parameters, so the goal criteria reach a minimum or maximum. Boundary conditions thereby describe the permitted range for the optimization characteristics.

As an example for a specific optimization tool which ensures a promising vehicle package in the early stages of the development process, the design tool Electric Vehicle Architecture Optimization Software (EVA-OS) is mentioned, illustrated in [21]. In this software, a closed model is established with mathematical relations between the vehicle's properties and the settings (characteristics) to be changed by the user. Thereby, a set of parameters represents a vehicle design and defines its properties. In order to calculate an optimization model of an overall vehicle architecture in a short period of time, the scope of characteristics and the level of detail are limited to a suitable degree. These measures reduce the calculation time for an optimization cycle, so it takes just a few hours. An impression of EVA-OS is illustrated in Figure 14.

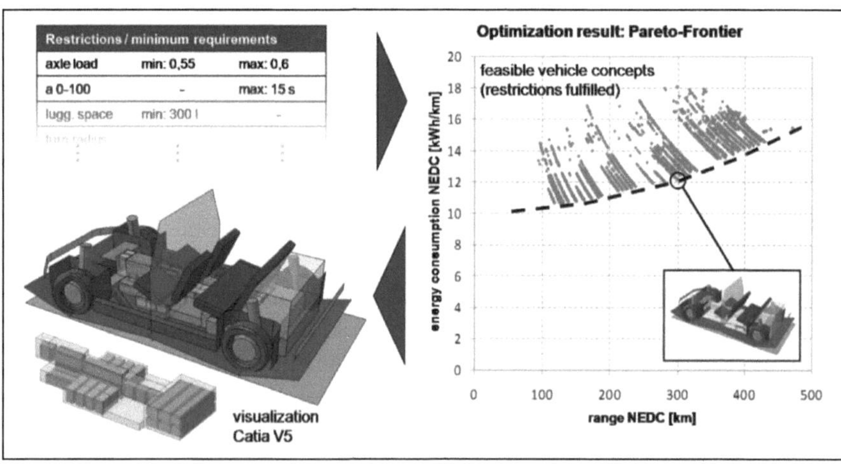

Figure 14: Process of vehicle layout with the use of an evolutionary algorithm in form of a 3D pareto frontier, [21]

10.4.3 Application of new technologies to solve goal conflicts

The creation of additional technical degrees of freedom by using new technologies is another possibility of handling goal conflicts. In this way, single characteristics can be changed and therefore properties of a concept can be influenced to solve the technical goal conflict. According to the example in Section 10.3.2, there are several possibilities of using other technologies to solve the goal conflict between the total weight and the drag coefficient regarding the vehicle length. If a relatively short vehicle is chosen, a smaller drag coefficient is hard to achieve by only changing the shape of the outer skin. Advanced technologies show a range from completely covering the under-floor via the demand-controlled opening and closing of air inlets in the vehicle front and the replacement of exterior mirrors by cameras up to actively influencing the air flow by outlets at certain positions. These measures can achieve significant drag coefficient improvements.

On the other hand, also the weight can be influenced in case of a longer vehicle. The substitution of material (e.g. light metal instead of steel), load-capable optimization of the design or the change of the body structure layout, including the application of fiber compound materials, are mentioned as examples of the spectrum of "new" technologies for weight reduction. In that case, the negative effect of increasing overall weight through a longer vehicle would be minimized or compensated, whereby an advantageous solution at least for this goal conflicts arises.

10.4.4 Vehicle variants to consider various sets of requirements

If various sets with spreads of requirements are identified during the concept phase of a vehicle, there has got to be a decision about the level of discrepancy regarding the spread and its consideration in the product development. One way to do this is to assemble significant spreads of requirements in a target system, [22]. As an example, such a relation system is illustrated in Figure 15 in a general way.

In the automotive industry, spreads of requirements within the target system are considered through various vehicle derivatives. To take into account as many spreads of requirements as possible, the implication would be a large number of several variants offered to the customer. On the other hand, there has to be a sufficiently large number of produced units to offer a vehicle with a competitive price to the customer. Therefore, the rate of consideration of spreads of requirements has to be estimated carefully when designing a new vehicle, [22].

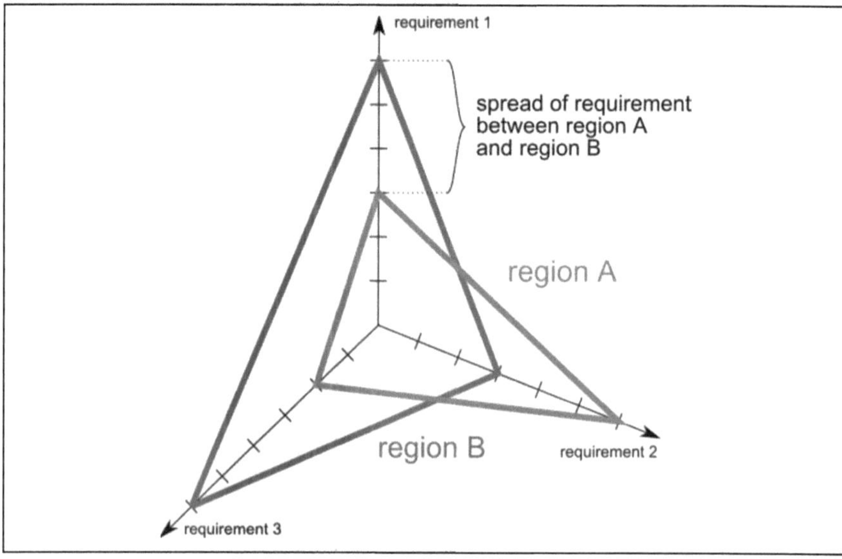

Figure 15: Illustration of varying solution spaces and the resulting spreads of requirements in a target system, referred to [22]

If no additional variants are developed, e.g. because of a small difference in a requirement, a compromise has to be found regarding the requirements on the vehicle. For this, the solution space of a requirement has to be analyzed, giving a range wherein the requirement is defined to be fulfilled. That range shows a compromise between the number of variants and the economical implementtation.

10.5 Conclusion

Because of an increasing globalization in the automotive sector, new target markets develop worldwide. Therefore, regional specific requirements have to be considered in the vehicle's development process. In order to meet these requirements and simultaneously minimizing the arising development time and costs, the design engineer has to be supported by specific methods and tools in the early stages of the development process. For clarifying and concretizing the task a model was presented, which systematically analyses the environment's influencing factors and impacts on the vehicle's technical characteristics and properties. The regional specific needs are represented by the sets of requirements, which have an influence on the technical parameters (characteristics and properties). These may be oppositional, so goal conflicts arise. In the target-oriented, methodologically supported management of requirements, properties, characteristics and goal conflicts, there is the potential to run an efficient product development, despite of increasingly diversified markets and high technical complexity. For advanced technical methods, e.g. modular development and assembly systems, requirements are hereby created already in the concept stage.

10.6 References

[1] Wallentowitz, H.; Freialdenhoven, A.; Olschewski, I. „Strategien in der Automobilindustrie – Technologietrends und Marktentwicklungen", Vieweg + Teubner, Wiesbaden, 2009.

[2] Nehuis, F.; Stechert, C.; Vietor, T. „Werkzeugunterstützung bei der Ermittlung von Anforderungen für regionsspezifische Fahrzeugkonzepte. In: 22. DfX-Symposium Conference Proceedings, 2011.

[3] Prinz, A. „Struktur und Ablaufmodell für das parametrische Entwerfen von Fahrzeugkonzepten", Logos Verlag, Berlin, 2011.

[4] Ponn, J. „Konzeptentwicklung und Gestaltung technischer Produkte – Systematisch von Anforderungen zu Konzepten und Gestaltlösungen", Springer Verlag, Berlin Heidelberg, 2011.

[5] VDI "Systematic approach to the development and design of technical systems and products", VDI-Gesellschaft Produkt- und Prozessgestaltung, Düsseldorf, 1993 (VDI-Richtlinie - Fachbereich Produktentwicklung und Mechatronik).

[6] Pahl, G.; Beitz, W. et al. „Konstruktionslehre, Methoden und Anwendungen", Springer-Verlag, Berlin, 2007.

[7] Stechert, C. „Modellierung komplexer Anforderungen", Dissertation TU Braunschweig, Verlag Dr. Hut, München, 2010.

[8] Ehrlenspiel, K. „Integrierte Produktentwicklung – Denkabläufe, Methodeneinsatz, Zusammenarbeit", 4. Aufl., Hanser, München, 2009.

[9] Weber, C. „How to Derive Application-Specific Design Methodologies", published in "Proceedings of the International Design Conference", Dubrovnik, 2008.

[10] Pahl, G. „Klären der Aufgabenstellung und Erarbeitung der Anforderungsliste", published in „Konstruktion Nr. 24", pp 195-199, Springer Verlag, Heidelberg, 1975.

[11] Franke, H.-J. „Methodische Schritte beim Klären konstruktiver Aufgabenstellungen“, published in „Konstruktion Nr. 27“, pp 395-402, Springer Verlag, Heidelberg, 1975.

[12] Kickermann, H. „Rechnerunterstützte Verarbeitung von Anforderungen im methodischen Konstruktionsprozess“, Dissertation TU Braunschweig, 1995.

[13] Franke, H.-J.; Grein, G.; Türck, E. „Anforderungsmanagement für kundenindividuelle Produkte – Methodische Lösungsansätze, Praxisorientierte Tools, Industrielle Anwendung“, Shaker Verlag, Aachen, 2011.

[14] Stechert, C.; Franke, H.-J. „Requirements Models for Collaborative Product Development”, published in “Proc. of the 19th CIRP Design Conference”, pp 24-31, 2009.

[15] Weilkiens, T. „Systems Engineering mit SysML/UML – Modellierung, Analyse Design“, dpunkt Verlag, Heidelberg, 2006.

[16] Fischer, H. „Nachhaltig führen lernen“, vdf Hochschulverlag AG an der ETH Zürich, 2004.

[17] Braess, U.; Seiffert, H.-H. „Vieweg Handbuch Kraftfahrzeugtechnik“, 5.edn., Vieweg Verlag, Wiesbaden, 2007.

[18] Ehrgott, M.; Nickel, S. „On the number of criteria needed to decide Pareto optimality“, published in “Mathematical Methods of Operations Research Volume 55”, 2000.

[19] Siebertz, K.; van Bebber, D.; Hochkirchen, T. „ Strategische Versuchsplanung – Design of Experiments (DoE)“, Springer Verlag, Heidelberg, 2010.

[20] Franke, H.-J. „Produktdarstellende Modelle als Hilfsmittel für eine weitergehende Rechnerunterstützung beim konzipierenden Entwerfen“ published in „7. Konstrukteurstag – Kammer der Technik“, edn. 1, Dresden, 1990.

[21] Kuchenbuch, K.; Vietor, T.; Stieg, J. “Optimisation Algorithms for the Design of Electric Cars”, ATZ worldwide, No. 8, pp. 16-19, 2011.

[22] Ehrlenspiel, K.; Kiewert, A.; Lindemann, U. „Cost-Efficient Design“, Springer Verlag, Berlin Heidelberg, 2007.

List of Authors

1

Richard Eckl, Peter Burda, Armin Förg, Harald Finke,
Prof. Dr.-Ing. Markus Lienkamp
Institute of Automotive Technology, Garching, Germany
E-mail: eckl@ftm.mw.tum.de

2

Prof. Dr. Andreas Daberkow, Marcus Ehlert and Dominik Kaiser
Dept. of Automotive Systems Engineering, Heilbronn University,
Heilbronn, Germany
E-mail:andreas.daberkow@hs-heilbronn.de

3

Bodo Gohla-Neudecker, Christian Kandler, Thomas Hamacher
Institute for Energy Economy and Application Technology
Technische Universität München, Germany
E-mail: B.Gohla-Neudecker@tum.de

4

Claus Mueller, Bernhard Hoffmann
Compact Dynamics GmbH, Starnberg, Germany
E-mail: claus.mueller@compact-dynamics.de

5

Prof. Dr.-Ing. Karsten Stahl, Dr.-Ing. Hermann Pflaum,
Dipl.-Ing. (FH) Georg Johann Meingaßner (lecturer)
Institute of Machine Elements – Gear Research Centre (FZG),
Technische Universität München (TUM), Garching, Germany

Prof. Dr.-Ing. Dr.-Ing. habil. Heinz Ulbrich, Dipl.-Ing. Johannes Mayer
Institute of Applied Mechanics, Technische Universität München (TUM),
Garching, Germany
E-mail: fzg@fzg.mw.tum.de

6

Prof. Dr.-Ing. Karsten Stahl, Prof. i. R. Dr.-Ing. Bernd-Robert Höhn,
Dr.-Ing. Hans-Philipp Otto, Dipl.-Ing. Richard Bäumler,
Dipl.-Ing. Christian Hasl (lecturer)
FZG Forschungsstelle für Zahnräder und Getriebebau, München, Germany
E-mail: ha@fzg.mw.tum.de

7

Andreas Kinigadner, **Eckhard Kirchner**, Rolf Najork
Schaeffler Technologies AG & Co. KG, Herzogenaurach, Germany
E-mail: eckhard.kirchner@schaeffler.com

8

Dipl.-Ing. Johannes Fuchs, Dipl.-Ing. Rainer Huber,
Prof. Dr.-Ing. Markus Lienkamp
Technische Universität München, Institute of Automotive Technology,
Garching, Germany
E-mail: fuchs@ftm.mw.tum.de

Dipl.-Ing. Thomas Riemenschneider
AUDI AG, Vehicle Concept Development, Ingolstadt, Germany

9

Dr. Alfred Pruckner, Stephan Kaspar, Dr. Ralf Stroph, Dr. Christoph Grote
BMW Research and Technology, München, Germany
E-mail: Alfred.Pruckner@bmw.de

10

Dr.-Ing. Alexander Prinz
Volkswagen Aktiengesellschaft, Wolfsburg, Germany
E-mail: alexander.prinz1@volkswagen.de

Dipl.-Ing. Frank Nehuis, Prof. Dr.-Ing. Thomas Vietor, Dr.-Ing. Carsten Stechert
Institute for Engineering Design, Braunschweig, Germany

Program Committee CoFAT 2012

Hans Adlkofer	Vice President, System Group Automotive, Infineon AG, München
Stefan Baier	Managing Partners & Shareholder, C-Con GmbH, München
Dr. Florian Bieberbach	Member of Management Board, Stadtwerke München GmbH, München
MdL Markus Blume	Member of the Bavarian Parliament, München
Maximilian Eck	Managing Director, Compact Dynamics GmbH, Starnberg
Rainer Eckl	Quality Director Europe, Analog Devices GmbH, München
Dr. Markus Eder	Bavarian State Chancellery, München
Prof. Dr.-Ing. Thomas Form	Head of Electronics Development, Vehicle Concepts, Volkswagen AG, Wolfsburg
Prof. Dr. Hubert Gasteiger	Chair of Technical Electrochemistry, (Dept. of Chemistry), TU München, Garching
Prof. Dr. Burkhard Göschel	Chief Technical Officer, Magna International, München
Dr. Günter Graf	Head of Department Industrial and Regional Policy, Bavarian Ministry of Economic Affairs, Infrastructure, Transport and Technology, München
Prof. Dr. rer. nat. Thomas Hamacher	Chair of Energy Economy and Application Technology, TU München, Garching
Prof. Dr.-Ing. Holger Hanselka	Director of Fraunhofer Institute for Structural Durability and System Reliability LBF, Darmstadt, Head of Institute System Reliability and Machine Acoustics SzM, TU Darmstadt, Darmstadt
Prof. Dr.-Ing. Hans-Georg Herzog	Specific field Energy Conversion Technology, TU München, Garching

Eberhard Hipp	Vice President, Central Division Research, MAN Truck & Bus AG, München
Prof. Dr.-Ing. Gerd Hirzinger	Head of the Robotics and Mechatronics Center, German Aerospace Center, Oberpfaffenhofen
Prof. Dr.-Ing Bernd-Robert Höhn	Chair of Machine Elements, TU München, Garching
Prof. Dr.-Ing Harry Hoster	Scientific Director Research Programme „CREATE", TU München, Garching
Christoph Huss	Vice President Development Abroad, BMW Group, München
Prof. Dr. Andreas Jossen	Chair of Energy Storage Technology, TU München, Garching
Prof. Dr.-Ing. Achim Kampker	Chair of Production Management, RWTH Aachen, Aachen
Prof. Dr.-Ing. Markus Lienkamp	Chair of Program Committee, Chair of Automotive Engineering, TU München, Garching
Steffen Lintz	Head of Branch Office Munich, IAV GmbH, München
Dr.-Ing. Christian Malorny	Principal, McKinsey & Company, Berlin
Franciscus van Meel	Director of Electrification Strategy, AUDI AG, Ingolstadt
Prof. Dr.-Ing. habil. Josef Nassauer	Chief Executive, Bayern Innovativ GmbH, Nuremberg
Bernd Neitzel	Head of the Hybrid & Electric Vehicle, Business Unit, Continental Engineering Services GmbH, Frankfurt
Prof. Dr.-Ing. Gunther Reinhart	Chair of Industrial Management and Assembly Technologies, TU München, Garching
Dr. rer. nat. Martin Sachenbacher	Emmy Noether Research Group Leader, TU München, Garching
Prof. Dr.-Ing. Ulrich Seiffert	CEO WiTech Engineering GmbH, Braunschweig
Kurt Sigl	President, Bundesverband eMobilität e.V., Berlin

Prof. Dr.-Ing. Gernot Spiegelberg	Head of Electro Mobility Siemens Corporate Technology, Siemens AG, München
Prof. Dr.-Ing. Karsten Stahl	Chair of Machine Elements, TU München, Garching
Dr. Rupert Stützle	Head of Systems Development for Electric Vehicles, Robert Bosch GmbH, Stuttgart
Prof. Dr.-Ing. Wolfram Volk	Chair of Metal Forming and Casting, TU München, Garching
Prof. Dr. rer. pol. Florian von Wangenheim	Chair of Services and Technology Marketing, TU München
Dr.-Ing. Lothar Wech	Business Development, TÜV SÜD Automotive GmbH, München
Karl-Friedrich Wörsdörfer	Vice President Product Development Hydraulic Brake Systems, Continental Teves AG & Co. oHG, Frankfurt
Prof. Dr.-Ing. Gebhard Wulfhorst	Chair of Land Use Patterns und Traffic Planning, TU München, Garching

MIX
Papier aus verantwortungsvollen Quellen
Paper from responsible sources
FSC® C105338

If you have any concerns about our products,
you can contact us on
ProductSafety@springernature.com

In case Publisher is established outside the EU,
the EU authorized representative is:
Springer Nature Customer Service Center GmbH
Europaplatz 3, 69115 Heidelberg, Germany

Printed by Libri Plureos GmbH
in Hamburg, Germany